Wildfires

Principles, Management Strategies, and Best Practices

Ankur Awadhiya

CRC Press
Taylor & Francis Group
Boca Raton London New York

CRC Press is an imprint of the
Taylor & Francis Group, an **informa** business

Designed cover image: Ankur Awadhiya

First edition published 2025
by CRC Press
2385 NW Executive Center Drive, Suite 320, Boca Raton FL 33431

and by CRC Press
4 Park Square, Milton Park, Abingdon, Oxon, OX14 4RN

CRC Press is an imprint of Taylor & Francis Group, LLC

ISBN: 978-1-032-83172-5 (hbk)
ISBN: 978-1-032-84636-1 (pbk)
ISBN: 978-1-003-51422-0 (ebk)

DOI: 10.1201/9781003514220

Publisher's note
This book has been prepared from camera-ready copy provided by the author.

To Ushi

Contents

SECTION IV Appendix

List of Figures

List of Tables

Preface

Wildfires are a menace — they not only decimate habitats and contribute to global warming, but also have long-lasting inimical effects on the local communities. They need to be managed assiduously, a lesson we've learnt profoundly and acutely during the Yellowstone National Park fires of 1988.

I became a nodal officer for forest fires in 2021 after Madhya Pradesh received 54,734 fire alerts in one fire season. Madhya Pradesh is an enormous state, with 52 districts (counties) and over 80 million people. The forests here store upwards of 2 billion tonnes of carbon dioxide, and are home to majestic animals including tigers, leopards, cheetahs, elephants, crocodiles, and vultures. All of these are at stake when a wildfire runs amok.

We made heavy use of data analytics in our strategy. We identified the times and places where wildfires occur, and we established the reasons for their occurrence. We deployed near-real time wildfire detection — SimplyFire — throughout the state, and generated mechanisms for quick transmission of information and resources to tackle wildfires.

The result was stupendous — wildfire alerts plummeted from 54,734 in the 2020–21 season to 34,559 in the 2021–22 season, and 16,647 in the 2022–23 season. The work was commended at national level, and I received a certificate of appreciation from the Head of Forest Force.

But throughout this process, two facts became glaringly evident — one, that wildfires can easily be tackled using data-driven strategies, and two, that documentation in this field is sorrowfully inadequate. Thus came the idea of documenting our tested approaches in the form of a book. I've tried to explain the concepts and principles, as I've understood through my academic studies, observations and learnings from practical experiences in the field.

I wish to thank the officers of Madhya Pradesh Forest Department who were kind enough to share their time and efforts in tackling wildfires during my tenure. My mother, despite her illness and old age, meticulously scrutinized the manuscript, and provided several constructive suggestions. Words are inadequate to express my profound sense of gratitude to her and to my family. Thanks are also due to NASA, ESA, OSM, OpenTopoMap, Google Earth Engine, R, and geemap teams for making visualizations and analyses possible.

Dr. Ankur Awadhiya, IFS

About the author

Dr. Ankur Awadhiya (b. 1987) is an IFS officer of 2014 batch borne on the Madhya Pradesh cadre. Trained as an engineer, he earned his B. Tech in Biological Sciences and Bioengineering in 2009 from IIT Kanpur, followed by a Ph.D from IIT Kanpur in 2015, AIGNFA (Honours Diploma) from the Indira Gandhi National Forest Academy in 2016 and Post Graduate Diploma in Advanced Wildlife Management (Honours Diploma) from the Wildlife Institute of India in 2018. He has studied Computer Science, Artificial Intelligence and Data Science from Harvard, Data, Economics and Development Policy from MIT and Remote Sensing from the University of Alaska. He maintains a keen interest in biodiversity conservation, forest management, and research, and is a prolific writer with over 50 publications and 2 patents. He has been a recipient of several honors including the NTSE scholarship, KVPY fellowship, Vanyaprani Sanrakshan Puraskar, Shri P. Srinivas Memorial prize, K. P. Sagriya Shreshta Vaniki Puraskar, and the S. K. Seth Prize. Besides academics and research, he maintains a passion towards teaching, photography, painting, movie-making, and creative literary pursuits.

Section I

Wildfire principles

1 Wildfires: An introduction

On the evening of November 7, Paradise, California, seemed to be a picture of serenity. The skies were clear, and a crisp chill had settled in, a herald of the approaching winter. Nestled at the base of the Sierra Nevada, the town had been enduring a gradual drop in temperatures, with daily highs lingering below 25°C and lows already brushing 5°C. The gentle breeze, moving at a mere 6 kilometers per hour, added a refreshing coolness to the air. As the weary residents — worn out from their daily grind — began to drift into a peaceful slumber, they did so with little apprehension. After all, they were in Paradise.

But as dawn broke the following morning, an insidious shift had occurred. The once pristine skies had taken on a murky, foreboding hue, as if cloaked by a dark, oppressive shroud. The air was thick with an acrid stench, and the town had transformed overnight into a scene of creeping dread. When the residents awoke, they were met with the terrifying sight of a towering column of smoke rising a few miles to the north and ashes fluttering down into their front yards.

By 7:15 a.m., the emergency 911 lines were overwhelmed with frantic calls from anxious residents, only to be told that they were not in immediate danger, as the fire was still some distance away.

But within an hour, the situation had escalated dramatically. A mandatory evacuation order had already been issued for the entire town of Paradise.

The inferno had ignited around 6:30 a.m. due to an equipment malfunction — an incident involving the Pacific Gas and Electric Company (PG&E), the largest utility company in America at the time. The blaze, sparked seven miles from town and separated by a deep canyon, was swiftly met with a concerted firefighting effort. But despite the rapid response, the relentless flames continued to spread, consuming over 150,000 acres before they were finally subdued more than two weeks later.

The catastrophic fire had claimed 85 civilian lives, reduced 18,000 structures to ashes, and inflicted damage exceeding 16 billion dollars in 2018 terms (Johnson, 2022).

That was a wildfire — a cataclysmic wildfire.

1.1 WHAT IS A WILDFIRE?

A wildfire, also known as a bushfire or forest fire, is a powerful and uncontrolled blaze that occurs in natural environments such as grasslands, shrublands, or forests — essentially areas abundant with combustible vegetation. These fires are typically spontaneous and unpredictable, compounded by a complex interplay of natural factors. However, with time, ingenious humans have developed a range of techniques to better understand and manage these events. Advanced technologies, including sophisticated computer simulations and satellite monitoring, have been employed to predict their occurrence and to control their spread more effectively. Despite these advancements, wildfires, left unchecked, have a tendency to expand uncontrollably, driven by factors such as wind, topography, and the availability of fuel. The consequences of such fires can be catastrophic, leading to extensive damage, a significant loss of life, and profound long-term changes to both human communities and natural ecosystems.

It is crucial to distinguish between wildfires and controlled, deliberate burns, commonly known as prescribed burns. Although both types of fires consume combustible vegetation, their purposes and management are fundamentally different. Prescribed burns are intentionally set as a proactive measure to reduce the risk of larger, more destructive wildfires. These controlled burns help manage fuel loads, reduce the likelihood of severe fires, and maintain ecosystem health. We shall explore controlled burning as a strategy for wildfire management in section 3.3.2. In contrast, wildfires occur with or without human intervention and can rapidly escalate into major disasters.

The cover of this book vividly illustrates a wildfire in Central India. This particular fire began at the base of a hill and swiftly climbed upwards, consuming all the available dry vegetation along its path. The image captures the intense and dynamic nature of such fires as they advance through the landscape, highlighting the relentless force of the blaze as it transforms the environment.

With the widespread availability of satellite data, we now have sophisticated tools for detecting, monitoring, and studying wildfires, something that we will discuss later in this book. For now, figure 1.1 showcases the wildfire that devastated the city of Paradise. The images featured in this figure are sourced from the Harmonized Landsat Sentinel-2 (HLS) project, which utilizes a virtual constellation of Landsat 8, Landsat 9, Sentinel-2A, and Sentinel-2B satellites. This integrated dataset offers comprehensive global coverage with a high temporal resolution of every 2 to 3 days, enabling detailed and timely observations of wildfire events.

Satellite data can be visualized in two primary ways: true color composites [Figure 1.1a,c,e], which represent scenes as they would appear to the human eye, and false color composites [Figure 1.1b,d,f], which reveal information not visible to the naked eye. For instance, the false color composites in figure 1.1 use short-wave infrared in the red channel, near-infrared in the green channel, and visible green in the blue channel. This approach is based on the fact that ash and char reflect strongly in short-wave infrared wavelengths, while healthy vegetation reflects prominently in near-infrared wavelengths. Consequently, in the false color images, burnt areas appear red, and vegetation appears green. The use of short-wave infrared is particularly advantageous because it can penetrate through smoke, allowing for the clear

(a) TCC dated 2018-10-07.

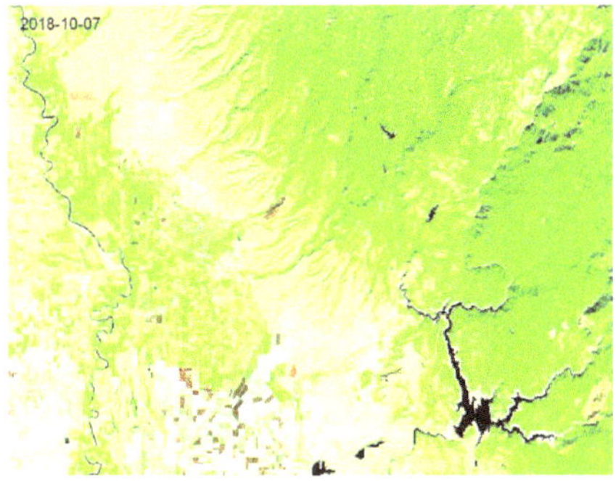

(b) FCC dated 2018-10-07.

(c) TCC dated 2018-11-08.

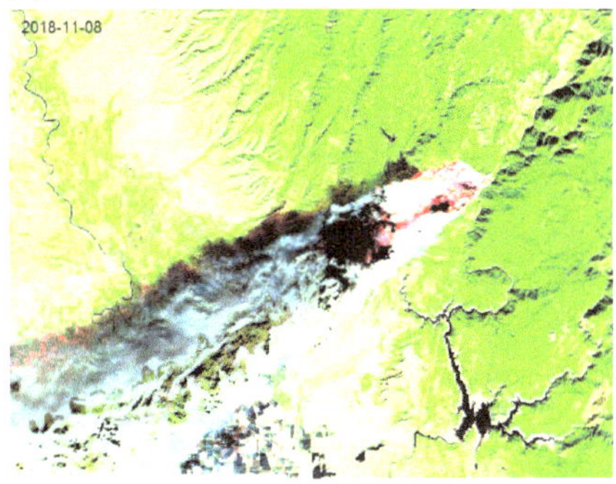

(d) FCC dated 2018-11-08.

(e) TCC dated 2018-12-26.

(f) FCC dated 2018-12-26.

FIGURE 1.1: True color composite (TCC) and false color composite (FCC) images of the Camp wildfire that caused wide-spread destruction in Paradise town.

identification of burnt areas even amidst dense smoke.

Analyzing figure 1.1, we observe several crucial details about the wildfire's impact. The pre-fire images (a and b) showcase a landscape rich in vegetation. The mountains are densely covered with greenery, and agricultural fields are prominently displayed as rectangular patches on the elevated terrains and also on the flatter areas. Some agricultural fields are marked in red in the false color composite images, indicating that these were subjected to burning — likely as part of a land-clearing process to remove agricultural residues and prepare the soil for new crops. As we will see later in this book, burning of agricultural wastes is an important reason for the initiation of wildfires throughout the world.

The images captured during the fire, dated November 8, 2018, present a dramatic scene. The fire's active front and the dense plume of smoke are clearly visible, moving predominantly in a South-West direction. On this particular day, strong winds, reaching speeds of 50 miles per hour, were blowing from the North-East towards the South-West. These intense winds played a critical role in exacerbating the wildfire's spread, driving the flames and smoke across the landscape at an accelerated rate. The combination of the fire's intensity and the powerful winds significantly influenced the wildfire's trajectory and the extent of its devastation. Details of wind movements are discussed in the appendix to this book.

1.2 DISTRIBUTION OF WILDFIRES

Wildfires are common on every continent except Antarctica [Figure 1.2a]. The map is constructed from satellite-derived data provided by NASA's Fire Information for Resource Management System (FIRMS), available at https://firms.modaps.eosdis.nasa.gov/. This comprehensive portal offers both near-real-time and historical fire data, utilizing advanced satellite instruments to monitor fire activity across the globe.

The data is gathered from several key sources: the Moderate Resolution Imaging Spectroradiometer (MODIS) instruments aboard NASA's Aqua and Terra satellites, and the Visible Infrared Imaging Radiometer Suite (VIIRS) on the Suomi National Polar-orbiting Partnership (S-NPP), NOAA-20, and NOAA-21 satellites. These instruments are crucial for fire monitoring due to their ability to capture detailed observations of the Earth's surface.

Fire detection is achieved through sophisticated contextual algorithms that analyze radiant energy across various wavelengths (Davies et al., 2008). These algorithms are designed to detect fires by evaluating the energy coming from the Earth's surface towards the satellite. For a fire to be detected, it must release substantial quantities of radiant energy across multiple wavelengths. Additionally, the fire must be brighter than its surrounding pixels to be distinguished from other features. The algorithms incorporate contextual threshold tests to ensure accurate identification, taking into account not just the raw energy levels but also the relative brightness compared to the background, which helps in distinguishing true fire signals

(a) Spatial distribution of fires as discerned from near-real time satellite alert data.

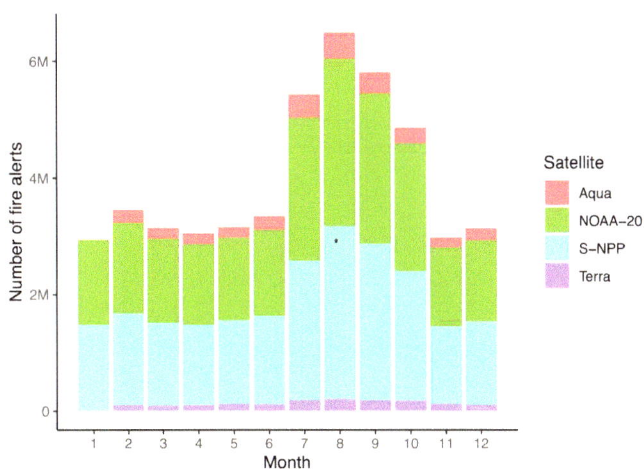

(b) Temporal distribution of fires as discerned from near-real time satellite alert data.

FIGURE 1.2: Spatial and temporal distribution of fires on Earth in year 2023.

from other sources of heat or light. By leveraging these advanced technologies and methodologies, FIRMS provides valuable information that aids in the management and response to wildfire incidents worldwide, enhancing our understanding and ability to address these critical events.

Figure 1.2a reveals that fires are notably rare in extremely cold regions such as Alaska, the northern parts of Canada, Greenland, Tibet, Mongolia, and much of Europe and Asia above 60°N latitude. This scarcity can be attributed to the harsh climatic conditions and permafrost in these areas, which result in very low vegetation (and fuel) density, and also inhibit the spread of fires. Similarly, areas below 60°S latitude in the Southern hemisphere are covered with water or ice, and do not have fire alerts. Fires are also uncommon in arid regions with limited vegetation, such as the Sahara Desert, the Thar Desert, and the Great Australian Desert. These regions lack sufficient plant matter to fuel fires, resulting in minimal fire activity.

The temporal distribution of fires, as depicted in figure 1.2b, plots the number of fire alerts detected by various satellites over different months of the year. This analysis indicates that fire incidents are recorded throughout the entire year. However, there is a noticeable spike in fire detections during the months of July, August, September, and October. These months consistently show a significantly higher frequency of fire alerts on a global scale, likely due to favorable conditions for fire spread, such as warmer temperatures and dry vegetation, especially in the Northern Hemisphere — that has more landmass than the Southern Hemisphere, and so affects any global-level analysis.

The spatial-temporal distribution of fires varies significantly across different regions, in response to the unique environmental and climatic conditions of each area. This variation underscores the importance of understanding local conditions when analyzing fire data. By examining these distributions, we can gain insights into the potential causes and patterns of fire incidents, which is essential for effective fire prevention and management strategies. Additionally, this information helps in prioritizing resources and planning fire-fighting operations more efficiently. Throughout this book, we will delve into the spatio-temporal distributions of fires in greater detail, highlighting their implications for fire management and safety.

1.3 TYPES AND CAUSES OF WILDFIRES

Wildfires are classified in various ways based on different criteria. Understanding these classifications is crucial for developing effective fire management strategies and mitigation efforts. Below is an overview of these classifications:

1.3.1 CLASSIFICATION BASED ON THE PARTS OF FORESTS CONSUMED BY THE WILDFIRE

1. **Surface fires**: These fires burn vegetation on the forest floor, including fallen leaves, twigs, grasses, debris, and other plant matter. These are com-

mon in deciduous forests where trees shed their leaves in drier months to conserve moisture. Surface fires typically burn at lower intensities since dead leaves and other surface materials have lower calorific values. Thus, these fires tend to be less aggressive, slowly moving, and easier to control. Surface fires can play a crucial ecological role by promoting seed germination for certain plant species and eliminating competing vegetation. They can also be extremely devastating for animals that take shelter in the fallen leaves, and for those that reside in the upper layers of soil (Certini et al., 2021).

2. **Ladder fires**: These fires start near the ground but spread to vegetation that is elevated above the ground but is below the tree canopy. This material can include vines, climbing plants, fallen branches, bark, shrubs, and small trees. Ladder fires have the potential to escalate wildfires from surface fires to canopy fires by providing a vertical path for flames to reach higher vegetation.

3. **Crown fires**: Also known as aerial or canopy fires, these fires consume the uppermost parts of trees, including the canopy, vines, mosses, and epiphytes like orchids. Crown fires spread rapidly, jumping from treetop to treetop. Due to their height and intensity, they often require advanced suppression techniques, such as aerial firefighting with helicopters or heavy-duty pumps. Crown fires can cause extensive damage to wildlife habitats and forests, resulting in significant ecological and economic losses.

4. **Ground fires**: Better termed "underground fires," these occur in areas with thick layers of organic matter, such as peatlands or locations with substantial leaf litter. Ground fires burn below the surface, making them challenging to detect. Signs of ground fires include heated ground and sudden combustion of snag trees[1]. These fires often produce a lot of smoke due to incomplete combustion occurring in oxygen-deprived situations.

1.3.2 CLASSIFICATION BASED ON ENERGY RELEASED BY THE WILDFIRE

1. **Low-intensity wildfires**: These fires release minimal energy per unit area per unit time. They typically occur in damp or wet conditions, with low fuel density or calorific value, or in cool, still, or insufficient air. Low-intensity wildfires generally have a lesser impact and are depicted as low burn severity areas on burn severity maps. We will discuss burn severity maps in detail in chapter 9. Since they are not highly destructive, these fires are occasionally created, managed, and allowed to burn by natural resource man-

[1]Snag trees are dead or dying trees, extremely important for wildlife as nesting sites.

agers. This controlled burning reduces the fuel load in the area, thereby helping to prevent more severe, high-intensity wildfires.

2. **High-intensity wildfires**: These fires release substantial amounts of energy per unit area per unit time. They often occur in dry conditions with high fuel density or calorific value, and in windy, hot environments. High-intensity wildfires have a significant impact on the ecosystem and human communities. Wildfire management strives hard to prevent them from happening in the first place, and to control them swiftly should they occur. They are represented as high burn severity areas on burn severity maps.

3. **Moderate-intensity wildfires**: These fires release an intermediate amount of energy between low and high-intensity fires. They are depicted as moderate burn severity areas on burn severity maps.

1.3.3 CLASSIFICATION BASED ON LOCATION AND FUEL TYPE

Grass fires are fires that occur in grasslands or prairies, where the primary fuel is grass. These fires can spread rapidly due to the abundance of dry grass, and they often play a role in maintaining the health of grassland ecosystems by promoting the growth of new vegetation and preventing the encroachment of woody plants.

In southern Africa, the term **veld fires** is used to describe fires that occur in veld, which are open grasslands. Veld fires are crucial for the health of these ecosystems as they help to recycle nutrients back into the soil, control plant growth, and support the survival of various animal species adapted to fire-prone environments.

Bush fires, on the other hand, take place in bushland or scrub areas. These fires typically involve a mix of shrubs and small trees, and their intensity can vary depending on the density of the vegetation. Bush fires are common in many regions and can be a natural part of the ecosystem, helping to clear out old growth and provide space for new plant life to thrive.

Forest fires occur in wooded, forested areas and can be particularly devastating as they consume not only trees but also underbrush and other forest vegetation. These fires can have significant impacts on wildlife habitats, air quality, and soil health. However, they also play a critical role in forest regeneration by clearing out old trees and allowing new growth to take place.

Hill fires are those that occur on hilly or mountainous terrains. The slope of the land can greatly influence fire behavior, making these fires challenging to manage. Inclined slopes can cause fires to spread quickly, and the uneven terrain can complicate efforts to control and extinguish them.

Desert fires occur in desert regions where vegetation is sparse. Although less frequent than fires in more vegetated areas, desert fires can still be significant, impacting the delicate balance of these arid ecosystems and affecting the plants and animals that have adapted to survive in such harsh conditions. Since deserts have scanty vegetation, any further loss of vegetation is especially detrimental to the desert ecosystem. For this reason, desert fires often have very long-term damaging effects.

Peat fires take place in peatlands, which are wetland areas where the fuel consists of accumulated peat, a type of organic soil formed from partially decayed plant material. Peat fires can be particularly challenging to extinguish because they often smolder beneath the surface, leading to prolonged and difficult-to-control burning.

Finally, **slash pile fires** are fires intentionally set in piles of cut branches, logs, and other debris that result from land clearing or forestry activities. These fires are often used as a method for managing and disposing of surplus vegetation, and while they help clear the land, they need to be carefully managed to prevent them from becoming uncontrolled and spreading to surrounding areas.

1.3.4 CLASSIFICATION BASED ON THE ORIGIN OF WILDFIRE

1. **Natural-origin wildfires**: These wildfires start due to natural causes. Common origins include lightning strikes, which can ignite dry vegetation, especially after prolonged dry periods or droughts. Volcanic activity, such as lava flows or hot ash, can also start fires in areas with abundant vegetation. Spontaneous combustion of organic matter, such as decomposing vegetation, can lead to fires, particularly in areas with high organic content. This is often facilitated by highly combustible gases like methane, created during the process of decomposition. Meteorite impacts can also trigger fires through the release of significant amounts of energy. We can note here that any natural-origin wildfire requires a very specific set of uncommon conditions, making them quite rare phenomena. For instance, lightning strikes are often accompanied by rainfall — wetting fuels and making them difficult to burn, and volcanism, spontaneous combustion, and meteorite impacts are themselves extremely infrequent occurrences. It is estimated that only around 10 to 30% of all wildfires globally are of natural origin (Robinne and Secretariat, 2021).

2. **Human-origin wildfires**: The majority of wildfires today — around 70 to 90% (Robinne and Secretariat, 2021) — are caused by human activities, which can be further categorized into intentional and accidental wildfires.

    - **Intentional wildfires**: These fires are intentionally set for a range of purposes, each aimed at achieving specific goals. One major reason is clearing of land for agriculture, which includes traditional slash-and-burn techniques as well as massive deforestation for modern industrial agriculture. In the latter case, large areas of forest are cleared to make way for expansive crop fields or pastures, often driven by industrial-scale farming operations.

      In some regions, fire is also used to stimulate the growth of new vegetation, which benefits livestock by providing fresh grazing opportunities. Additionally, burns are employed in the collection of forest products, such as berries or medicinal plants, helping clear underbrush and improve accessibility inside forests.

Fires are sometimes employed for more unconventional purposes as well. They may be used to drive animals into traps for hunting or capture, or to conceal evidence of criminal activities such as large-scale fellings. In the digital age, some individuals start fires to create sensational content for social media, seeking attention and engagement. We also have instances of revenge wildfires where offenders initiate fires to strike vengeance or retribution upon government officials. Some communities also have a tradition of creating wildfires as a tribute to Gods for accepting their prayers, especially after a good harvest.

Regardless of intent, these deliberate fires carry substantial risks. If not managed properly, they can quickly spiral out of control, evolving into large-scale wildfires that cause extensive environmental damage, threaten wildlife, and endanger human lives and property.

- **Accidental wildfires**: These fires often result from negligence or unintended causes rather than deliberate actions. Common examples include unattended campfires, which can easily spread when left without supervision. Discarded cigarettes can ignite dry vegetation, leading to wildfires when they are not properly extinguished. Sparks from equipment, such as power lines or chainsaws, are another potential hazard, as these sparks can ignite surrounding materials. We have seen before that the Camp wildfire that caused widespread destruction in Paradise town had started due to an equipment malfunction.

  Vehicles, especially ill-maintained ones, may also initiate wildfires if they create sparks or when their hot components come proximate to tall, dry grasses. Fireworks, when used irresponsibly or set off in dry conditions, also pose significant risks of starting fires.

  The disposal of debris through burning — especially during windy conditions — can also result in wildfires, as strong winds can carry embers to new areas and ignite additional fires. This is especially common in modern agricultural practices involving burning crop residues to clear fields. In all these cases, lack of proper precautions and control measures can turn minor incidents into major wildfires, causing extensive damage to the environment, wildlife, and human communities.

1.4 THE FIRE TRIANGLE

The fire triangle, also known as the combustion triangle, is a fundamental concept used to understand the conditions necessary for a fire to ignite and sustain itself

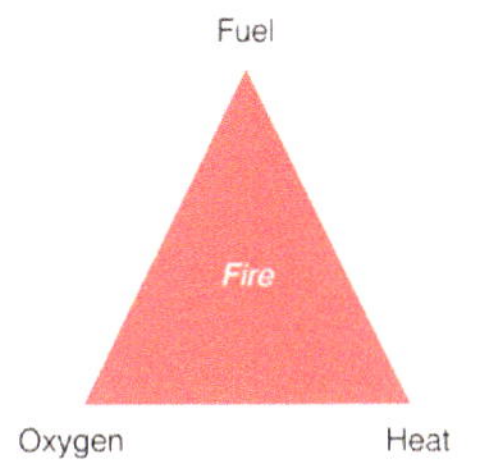

(a) The fire triangle.

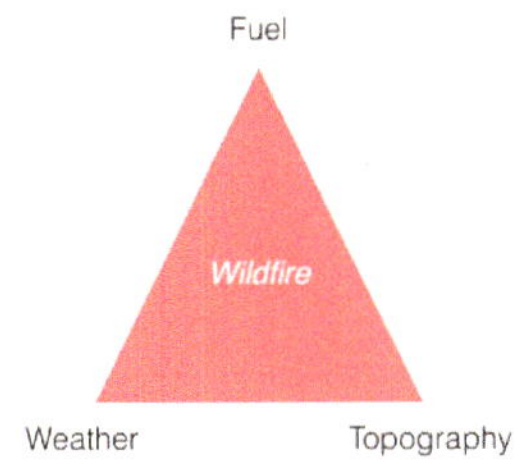

(b) The wildfire triangle.

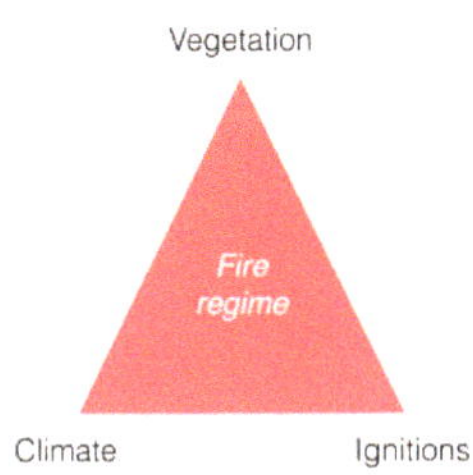

(c) The fire regime triangle.

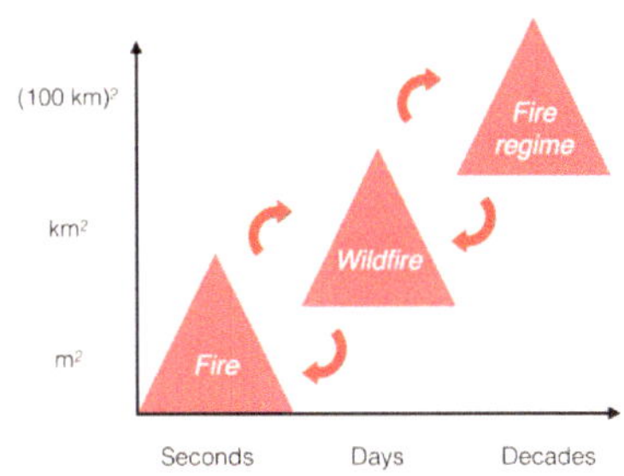

(d) Interactions between different fire triangles.

FIGURE 1.3: Fire triangles at different spatial and temporal scales.

[Figure 1.3a]. It illustrates that three key elements are required for a fire: fuel, oxygen, and heat. Fuel is any combustible material that can burn, oxygen supports the combustion process, and heat provides the energy needed to start and maintain the fire. If any one of these elements is removed, the fire will be unable to continue and will eventually be extinguished. Understanding this concept is crucial — both for preventing, and also for controlling fires effectively.

The first essential condition for fire is fuel, which refers to any material capable of reacting with oxygen (or other chemicals) to release energy. In the context of wildfires, this fuel primarily consists of organic materials such as leaf litter, fallen twigs and branches, soil humus, and various forms of vegetation including trees, vines, and climbing plants. These materials contain chemical energy stored in the form of organic molecules; this energy was originally captured during photosynthesis — the process by which plants convert sunlight into chemical energy. When the fuel burns, it releases this stored energy in the form of heat and light. The burning process occurs in three stages:

1. **Preheating phase:** During this initial stage, the unburnt fuel is heated, leading to the release of ignitable vapors through processes such as dry dis-

tillation — involving thermolysis[2], pyrolysis[3], and cracking[4]. As the temperature rises, it reaches the flash point of the fuel, the lowest temperature at which ignitable vapors are emitted in sufficient quantities to form a flammable mixture with air. The temperature then climbs to the fire point, also known as the combustion point, where the vapors continue to burn for several seconds after ignition.

2. **Gaseous or distillation phase:** In this stage, the ignitable vapor-air mixture burns after reaching its combustion point, resulting in a significant release of heat and light. Flames are typically visible as the gas mixture burns. The generated heat spreads to nearby materials through conduction, convection, and radiation, bringing them to their preheating phase and priming them to ignite. This creates a chain reaction that allows the fire to persist and spread.

3. **Charcoal or solid phase:** As fire progresses, the release of ignitable gases diminishes, and flames become less prominent. The remaining fuel is reduced to charred remains, which consist of carbon and unburned minerals. This char continues to burn in a slow, flameless combustion process known as smoldering. Smoldering produces high temperatures and can persist for an extended period, as the process is slow and the generated heat is retained within the char.

The second condition necessary for combustion is oxygen. Oxygen acts as an oxidizing agent in the combustion reaction. This gas constitutes about 21% of the Earth's atmosphere in mole fraction, and around 23% of the Earth's atmosphere in mass fraction. Most atmospheric oxygen is produced by plants through photosynthesis, a process summarized by the chemical equation:

$$6CO_2 + 6H_2O \rightarrow C_6H_{12}O_6 + 6O_2$$

While other substances like halogens (e.g., fluorine, chlorine), ozone, and sulfur oxides can also serve as oxidizing agents, oxygen is by far the most common and vital for wildfires.

The third condition is heat, which provides the activation energy required to initiate the combustion reaction. Since combustion is an exothermic process (releasing heat), once the reaction begins, it can sustain itself and even accelerate. This self-perpetuating process can lead to a chain reaction, where the heat generated from the initial combustion ignites additional fuel, causing the fire to spread.

---

[2] Thermolysis is the breakdown of molecules by the action of heat.

[3] Pyrolysis is decomposition brought about by high temperatures.

[4] Cracking is the process of breaking down of complex organic molecules into simpler molecules by the breakage of carbon-carbon bonds.

Understanding the fire triangle — fuel, oxygen, and heat — provides crucial insights for fire control strategies. Since all three conditions are necessary for a fire to burn, removing any one of them can effectively extinguish a fire. For example, fire lines are created by clearing a strip of vegetation to remove fuel, preventing the fire from spreading beyond this break. In smaller fires, such as kitchen fires, using fire blankets, sand, or carbon dioxide can cut off the oxygen supply and extinguish the flames.

Reducing the temperature by dousing fires with water is another widely used technique, which is why fire hydrants are installed in cities to provide a ready supply of pressurized water for firefighting.

In addition to the classic fire triangle, we can also consider the wildfire triangle [Figure 1.3b] and the fire regime triangle [Figure 1.3c] at broader spatial and temporal scales (Moritz et al., 2005).

At the scale of wildfires, factors such as the fuel load, weather conditions, and topography play significant roles. For instance, hot and dry conditions can make fuels more susceptible to ignition, while wind can intensify an already burning fire. Topography influences fire behavior as well — fires on hilly terrains can spread rapidly with flames rising up, and also through convection currents — and slope-driven winds can exacerbate wildfires, as we saw in the case of the Camp wildfire.

Fire regimes analyze the broader patterns and characteristics of fires over large areas and extended time periods [Figure 1.3c]. This includes studying the frequency, intensity, and impact of wildfires on ecosystems. Fire regimes are assessed at scales of hundreds of square kilometers and several decades, focusing on vegetation types and climate patterns. Vegetation types can include species that depend on fire for their life cycles, and climate encompasses long-term averages of temperature, precipitation, and wind patterns occurring in the area.

These scales are interconnected, as illustrated by the curved arrows in figure 1.3d. Small fires can evolve into larger wildfires, and large wildfires can create numerous spot fires in their vicinity. Frequent wildfires can alter species composition, favoring fire-resistant or fire-dependent species, thereby changing the long-term fire regime of an area. Conversely, the prevailing fire regime influences the likelihood of wildfires at smaller scales, shaping fire behavior over time.

1.5 TRENDS IN WILDFIRES

Wildfire trends exhibit considerable variability depending on the region and time period under consideration. In the United States, the area burned by wildfires has shown a significant upward trend over the years [Figure 1.4a]. During the 1980s, approximately 2.5 million acres were burnt annually. In contrast, recent years have seen this number swell to between 5 million and 10 million acres per year. Additionally, data reveal fluctuations over time, with periods of increased burning often followed by lulls. This pattern is anticipated because wildfires consume the available fuel load — primarily vegetation — in forested areas. After large fires, it takes time for the vegetation to recover and reach pre-fire levels, which can lead to temporary decreases in the frequency and intensity of wildfires till the fuel load builds up again.

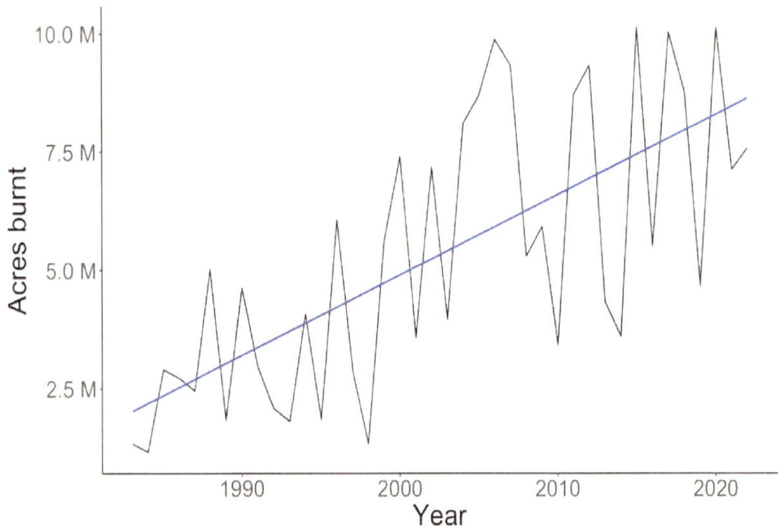

(a) Area burnt in US wildfires over time. Data source: National Interagency Fire Center https://www.nifc.gov/fire-information/statistics/wildfires

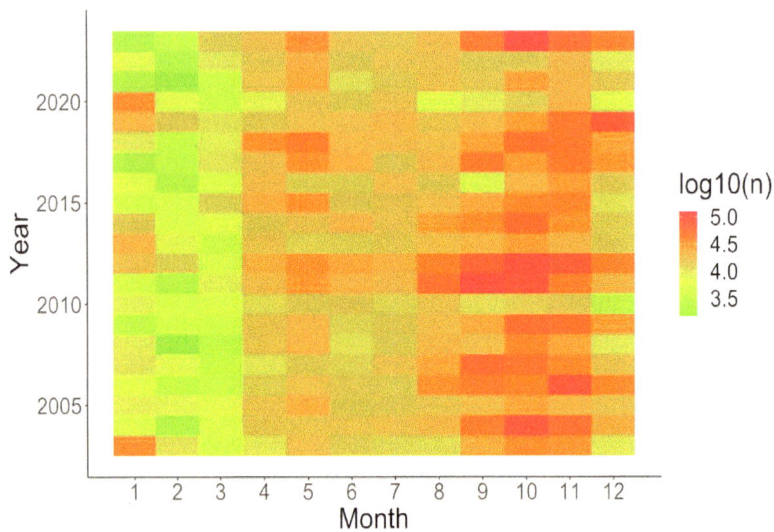

(b) Heat map of MODIS fire alerts from Terra and Aqua satellites over Australia in different months and years. Green represents few fire alerts, and red represents a large number of fire alerts. Data source: https://firms.modaps.eosdis.nasa.gov/download/list.php

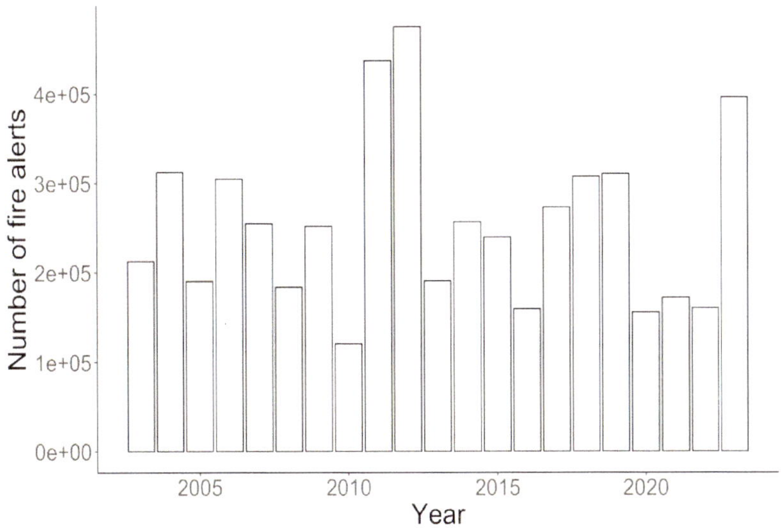

(c) Trend of number of MODIS fire alerts from Terra and Aqua satellites over Australia in different years. Data source: https://firms.modaps.eosdis.nasa.gov/download/list.php

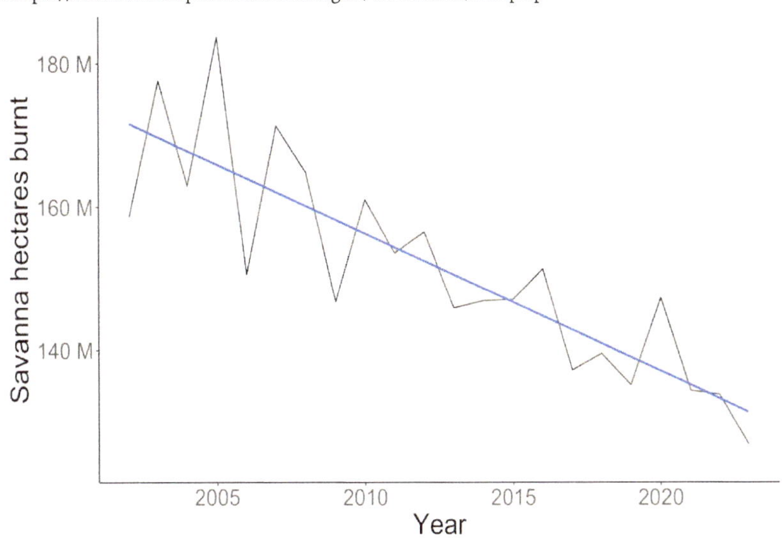

(d) Global savanna area burnt in different years. Data source: https://gwis.jrc.ec.europa.eu/apps/country.profile/downloads

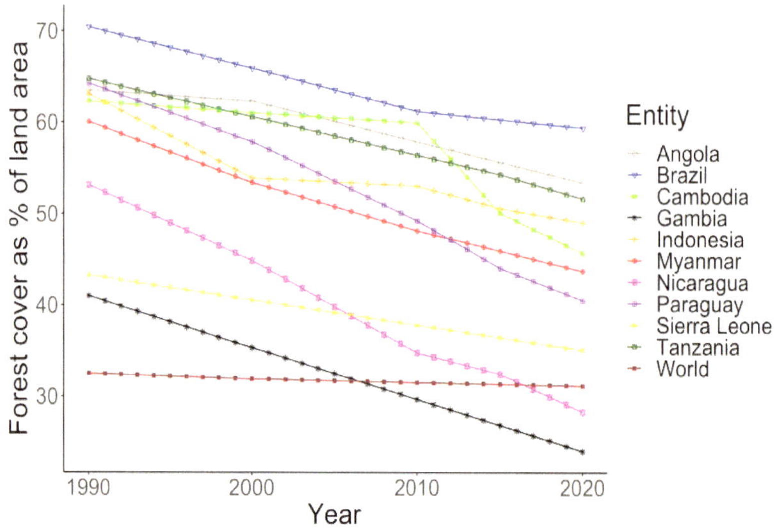

(e) Percentage of land area covered by forests in different countries and the world. Data source: https://fra-data.fao.org/ and https://ourworldindata.org/deforestation

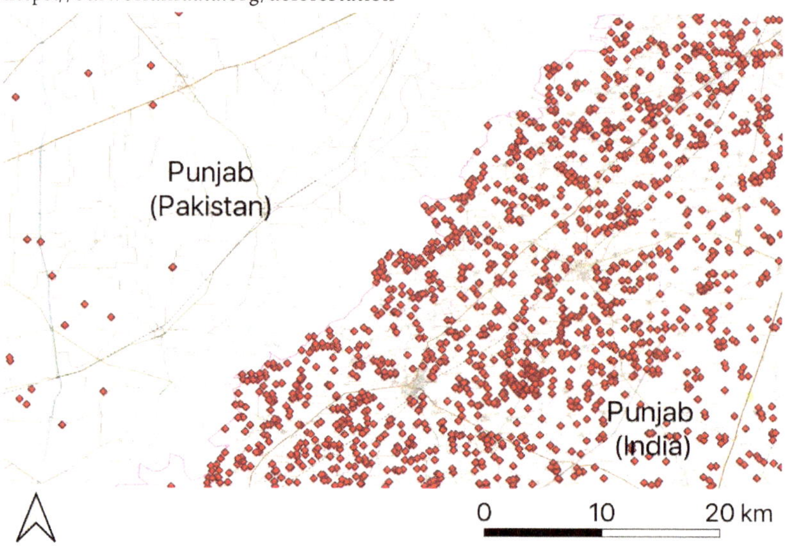

(f) Fire points in November 2023 around the India-Pakistan border (purple) in Punjab. Data source: https://firms.modaps.eosdis. nasa.gov/download/list.php

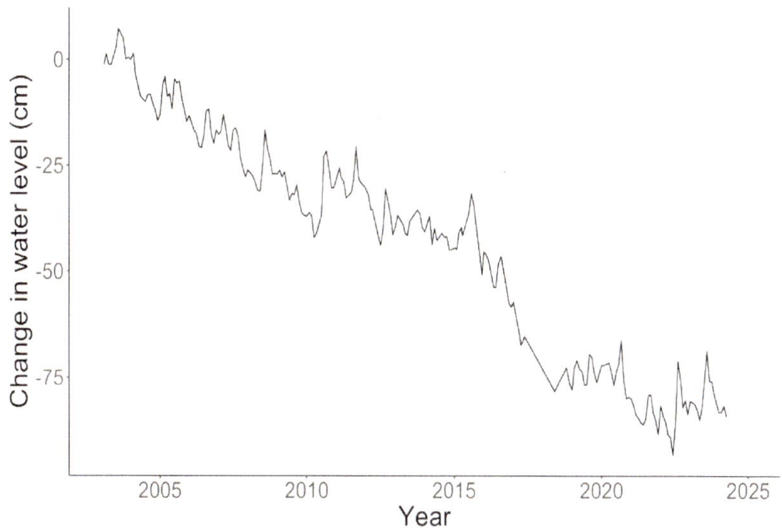

(g) Change in groundwater level in Ludhiana district of Punjab, India. Data source: https://grace.jpl.nasa.gov

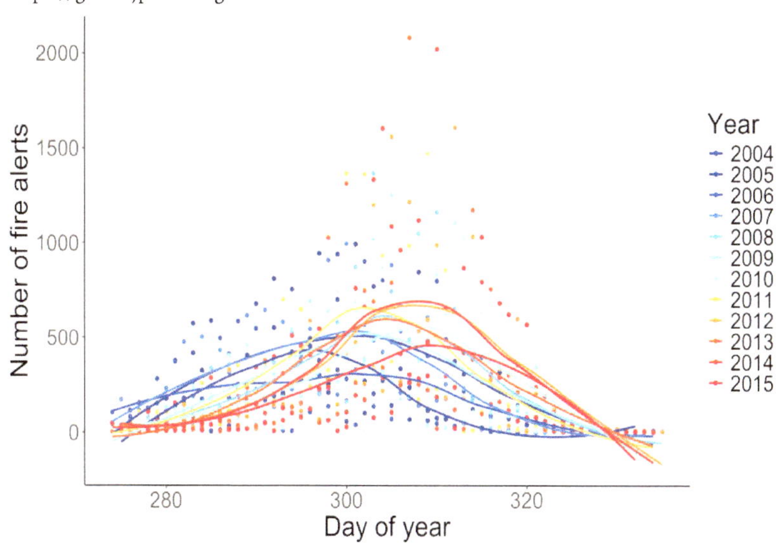

(h) Graph of daily fire alerts in Punjab, India over several years, with trendlines. Data source: https://firms.modaps.eosdis.nasa.gov/ download/list.php

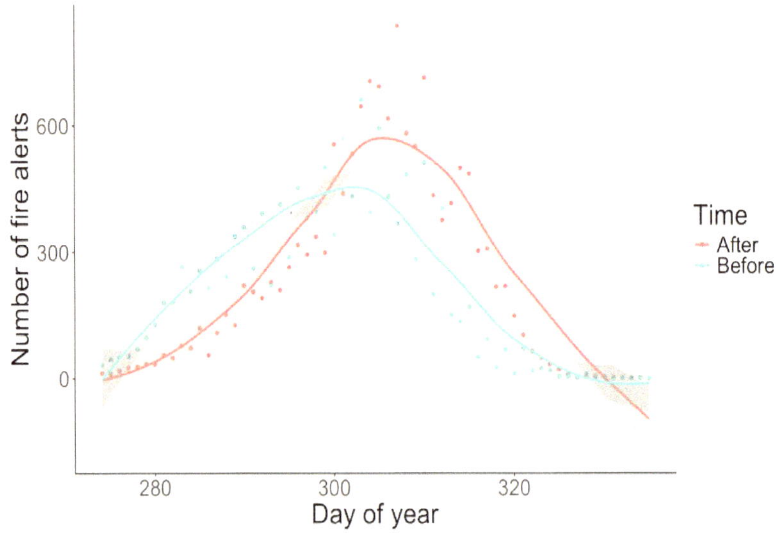

(i) Categorization of data in (h) into years before and after the passage of the Punjab Preservation of Subsoil Water Act 2009. Data source: https://firms.modaps.eosdis.nasa.gov/download/list.php

FIGURE 1.4: Trends in fire and forest cover.

The availability of satellite-based fire detection technologies has significantly improved our ability to collect and analyze wildfire data, enabling detailed examinations of fire patterns. However, it is important to recognize that satellites detect fires in various landscapes, not just in forest or bushland areas. Therefore, the fire data may also include fires in non-forest areas. Despite this, the extensive data we now have allows us to discern several key trends.

For example, when plotting fire alerts from the MODIS sensor as a heat map [Figure 1.4b], we see that Australia experiences the highest number of fire alerts during September, October, and November. In contrast, January, February, and March typically see fewer fires. Data for March over the years also reveals a trend — earlier years show more green shades, indicating fewer fire alerts, while recent years show more red shades, indicating an increase in fire alerts over time. Additionally, there is a noticeable variability in the data; for instance, September 2016 had unusually low fire alerts, while December 2019 had exceptionally high numbers. This variability highlights the need for such wildfire management plans that can account for these fluctuations, or in other words, adaptive management plans.

In examining yearly fire alerts in Australia [Figure 1.4c], the trends differ significantly from those observed in the United States [Figure 1.4a]. Australian data does not show a clear upward trend but rather a zig-zag pattern. After several years of high fire activity, there is often a period of lower fire alerts, followed by a resurgence. For example, 2008 had fewer fire alerts, partly due to greater fires in the previous years and also due to above-average rainfall that reduced fire risk. However, increased rainfall also stimulates vegetation growth, which can lead to higher fire risks in subsequent years. This was evident in 2009 when Victoria, having experienced few fires in 2008, faced severe wildfires during the Black Saturday bushfires, which ravaged 1.1 million acres and resulted in 173 fatalities (Cameron et al., 2009).

In 2010, there was an exceptionally low number of fire alerts [Figure 1.4c], due to effective prevention measures and a strong La Niña effect, which brought abundant rainfall and flooding to Queensland and Victoria. However, the following years again saw an increase in fire alerts as conditions became favorable for burning the accumulated vegetation.

Another trend is the decrease in global savanna area burnt [Figure 1.4d]. Savanna areas refer to landscapes where grasslands dominate, interspersed with scattered trees. These regions typically experience a warm climate with seasonal rainfall, allowing for rich biodiversity and unique ecosystems. Savannas are crucial for various animal species and often serve as important grazing grounds. Before 2005, over 170 million hectares of savanna were burnt annually, but current estimates are below 140 million hectares.

This decline is partly due to improved forecasting and management practices. But the trend is also reflective of a reduction in global savanna areas. In regions like Australia, savannas are being converted into grasslands for livestock grazing (Hoffmann and Jackson, 2000). Conversely, in many African countries, tree planting is converting savannas into woodlands (Armstrong et al., 1998; Parr et al., 2024). Both of these conversions have been leading to a reduction in the total savanna area.

In this way, what we observe in savanna fire trends is not just a reduction in fires due to better management, but also a transformation of savannas into other land uses — often due to poor management of resources, or poor protection of existing savanna areas permitting encroachment and conversion to other land uses, or both (Jepson, 2005). While the first reason is a cause of hope and joy, the second paints a dismal picture and necessitates urgent, extensive, and concerted efforts if natural ecosystems are to be preserved.

In causal inference studies, it is crucial to consider confounding variables that may obscure the true relationship between variables through spurious correlations. When we observe a decrease in wildfire incidents, this might be misinterpreted as a sign of improved fire management, when in reality, it could be due to reduced forest cover. Figure 1.4e shows a decline in forest cover across many countries. As forest areas shrink, the number of fire alerts may also decrease, which could misleadingly suggest that fire management practices are more effective than they actually are. The reduction in forest cover can be due to various factors, including deforestation for agriculture and urban development. Hence, a comprehensive examination of confounding variables is essential to accurately assess fire management strategies and ensure that the interventions taken are genuinely effective — not merely a response to reduced forest availability.

Fire trends also emanate from policy differences. Sharp boundaries in fire signatures, such as those captured by the VIIRS sensor in November 2023 near the India-Pakistan border [Figure 1.4f], reveal striking differences in fire activity across artificial boundaries. From the figure, we can observe that the province of Punjab in India shows a significantly higher density of fire signatures compared to its neighboring province in Pakistan. Despite these areas having a similar geography, shared culture and history — even to the extent of having a common name — Punjab, there is a marked difference in fires. This can be traced to man-made factors and differences in government policies on the two sides of the international border.

In India, the Punjab Preservation of Subsoil Water Act 2009 (PPSW Act 2009) was introduced to combat groundwater depletion, which had been exacerbated by resource-intensive agricultural practices. The region's intense agricultural activities, particularly paddy cultivation, had significantly strained groundwater resources, and the water table was going down year after year [Figure 1.4g]. To mitigate this, the Act regulated the timing of paddy planting, aiming to reduce groundwater extraction by aligning crop sowing with the monsoon season.

However, this policy also had several unintended consequences. The delay in planting paddy resulted in a delay in harvesting paddy. This means that the accumulated crop residues have to be cleared in a very short period of time to clear fields in preparation for the next planting season. To accomplish this, farmers have increasingly resorted to burning crop residues as a quick and cost-effective solution to clear their fields. It is this increase in burning that is clearly reflected in the satellite data. Fire signatures have been going up and moving towards the end of the year [Figure 1.4h–i]. In contrast, Pakistan, which does not have similar regulations, allows farmers more flexibility and time to manage their crop residues, and we observe fewer fire alerts in Pakistan. This example illustrates how policy changes can directly impact fire activity and its distribution, revealing a complex interaction between policies, local practices, and environmental protection.

1.6 WILDFIRES AND SOCIETY

Wildfires have deeply influenced human societies throughout history, shaping our evolution (Glikson, 2013) — and they continue to impact our modern lives even today. The earliest sources of fire for humans were likely wildfires, and they likely played a pivotal role in early human development (Gowlett, 2016). Early humans would have encountered and harnessed fire from natural wildfires, using it for cooking, protection, and hunting. This initial interaction with fire was crucial for survival and led to significant advancements in human society. Over time, managing and controlling fire became integral to human adaptation, influencing cultural practices, land use, and community structures.

The impacts of wildfires are extensive and interconnected even today, affecting economic, public health, and environmental domains. Economically, wildfires result in substantial costs, including firefighting efforts, property damage, and loss of natural resources, with additional indirect costs such as business disruptions and decline in tourism. Public health is severely impacted by wildfire smoke, which contains harmful pollutants that exacerbate respiratory conditions like asthma and chronic obstructive pulmonary disease (COPD), as well as by the stress associated with evacuations. Environmentally, wildfires alter landscapes, destroy habitats, contribute to soil erosion, and degrade water quality, while also negatively affecting biodiversity and disrupting ecological balance. These combined effects underscore the need for comprehensive wildfire management strategies that address their broad-ranging consequences. The next chapter will delve deeper into the various effects of wildfires.

Wildfire management has been undertaken since antiquity by societies to counter and abate the destructive and detrimental consequences of wildfires. The repertoire of traditional knowledge from indigenous and native societies provides us with several valuable insights into effective wildfire management. For instance, Aboriginal Australians use cultural burning to manage vegetation, reduce fuel loads, and support ecosystem health. Similarly, Native American tribes employ controlled burns to maintain forest ecosystems and foster species dependent on fire. Integrating these traditional practices with modern fire management techniques offers a more holistic approach to addressing wildfire challenges through the incorporation of time-tested methods.

At the same time, we need to keep evolving our methods. To address the increasing frequency and intensity of wildfires — due to climate change, and also because of increasing man-made causes — modern society must adapt by developing fire-resistant infrastructure and implementing proactive fire-management strategies. Advances in construction materials, fire-resistant landscaping, and strategic community planning are essential for enhancing resilience against fires. Public education on fire risks, evacuation procedures, and creating defensible spaces around homes is also crucial. By combining modern techniques with traditional practices and investing in resilient infrastructure, societies can better manage the challenges posed by wildfires and continue to adapt and thrive in the face of their persistent influence.

2 Effects of wildfires

Wildfires occur in environments with substantial fuel loads, primarily consisting of vegetation and organic matter. These areas often have dense forests and diverse wildlife, both of which are already under considerable stress from a variety of factors, many of which are man-made. In this way, wildfires significantly worsen the ongoing crisis of declining biodiversity by not only killing plants and animals, but also by decimating habitats and ecosystems.

Wildfires also result in considerable destruction to human communities. Besides loss of life and property, they also have long-term deleterious impacts on soil erosion, fertility, availability of fresh water, and tourism avenues. Furthermore, wildfires also emit harmful gases, smoke, and ash, which contribute to the expansion of their detrimental impact on the health of humans, plants, and animals.

Wildfires also have several short, medium, and long-term inimical effects on the society at large, causing disruptions in social life, evacuations, and forced migrations. Through the release of copious amounts of greenhouse gases, soot, and nucleation particles, wildfires are also changing weather and climate patterns over small and large scales.

In this chapter, we will look at these, and many more effects of wildfires in detail.

2.1 CLASSIFICATION OF WILDFIRE EFFECTS

The effects of wildfires can be systematically categorized into several classifications:

1. **Spatial and temporal effects**: We can examine effects of wildfires in space (spatial dimension), time (temporal dimension), or both at the same time.
   a. **Spatial effects**: The impacts of wildfires can be analyzed in terms of their geographical extent and distribution. Some effects are localized to the immediate vicinity of the fire, while others are spread over much larger areas. In this way, spatial effects are divided into proximate, intermediate, and distal effects:
      - **Proximate effects**: These are the direct impacts experienced in the immediate area surrounding the wildfire. The intense heat and flames can cause significant damage to vegetation, soil, and wildlife in this zone. We have seen this proximate area in figure 1.1e–f, where the burnt region is visible.
      - **Intermediate effects**: These occur at a moderate distance from the fire. While the intensity of the impact is less severe compared to the proximate area, effects such as smoke deposition and altered air quality can still be significant.

- **Distal effects**: These are observed at large distances from the wildfire. The effects in this zone primarily include the dispersal of smoke, ash, and other pollutants, which can affect air and water quality, and weather, over a large area. We have seen this distal area in figure 1.1c–d, where smoke emanating from the Camp wildfire is visible. A similar picture emerges when plotting distribution of various pollutants released during a wildfire, as compared to baseline values in years without comparable wildfires. For example, in figure 2.1, we observe very high concentrations of aerosols at the site of Camp wildfire, and a considerable concentration of aerosols even at locations that are significantly far off from the site of the wildfire, to which these aerosols would have reached through wind movement, diffusion, and other processes. We do not find a similar picture in other baseline years that did not have huge wildfires; though such years do depict some distribution of aerosols due to other reasons. In this context, the concentration of aerosols is measured using the Ultraviolet Aerosol Index (UVAI), a quantitative measure of aerosol concentration in the atmosphere, derived from the difference in ultraviolet radiation intensity under aerosol-laden conditions compared to clear sky conditions. The formula for calculating UVAI is: $UVAI = \frac{(I_{UV} - I_{UV,\text{clear}})}{I_{UV,\text{clear}}}$, where I_{UV} represents the measured ultraviolet radiation intensity at a specific wavelength in the presence of aerosols, while $I_{UV,\text{clear}}$ denotes the expected intensity of UV radiation in a clear atmosphere devoid of aerosols. The resulting UVAI provides insight into the concentration of aerosols, with higher UVAI values indicating a greater abundance of aerosols. This facilitates the assessment of aerosol impacts on atmospheric conditions, weather, and climate.

b. **Temporal effects**: The consequences of wildfires can also be assessed over time, considering both the immediate aftermath and the longer-term changes that result from the fire. Temporal effects are classified into short-term, medium-term, and long-term effects:
   - **Short-term effects**: These refer to the immediate impacts of wildfires, such as the destruction of vegetation and the disruption of local ecosystems. For instance, grasses may be destroyed, but they can often regenerate relatively quickly with subsequent rainfall.
   - **Medium-term effects**: These include the changes that occur over a period of months to years following the wildfire. This may involve shifts in species composition, changes in soil properties, and the gradual recovery of vegetation.
   - **Long-term effects**: These are the enduring consequences that can persist for decades or even longer, as seen in figure 2.2 (NASA, 2019). Such effects include the prolonged loss of biodiversity, alterations in ecosys-

(a) Location of the city of Paradise on topographical map of West Coast of United States.

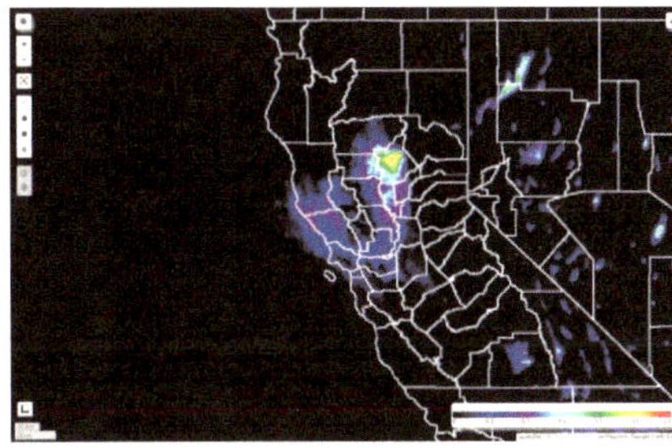

(b) Distribution of aerosols in November 2018 at the time of Camp wildfire, revealing greatest concentration at the site of the wildfire.

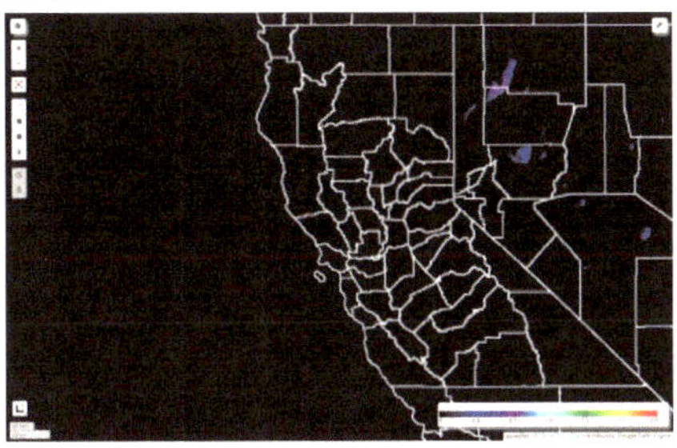

(c) Baseline distribution of aerosols in November 2019, a year after the Camp wildfire, for comparison.

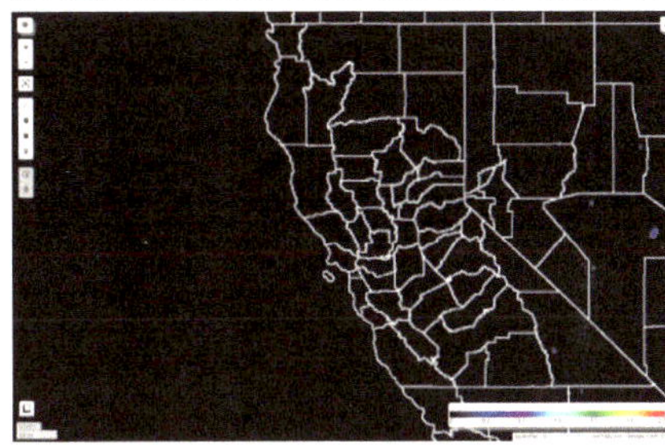

(d) Baseline distribution of aerosols in November 2020, two years after the Camp wildfire, for comparison.

FIGURE 2.1: Distribution of aerosols around Camp wildfire as discerned through Sentinel-5P UV aerosol index. Data courtesy European Space Agency, Open Street Map, and Open Topo Map. Visualized on geemap using Google Earth Engine.

tem structure and function, and the potential for permanent extinction of species.

c. **Combination of spatial and temporal effects**: Each combination provides insights into how the impacts of wildfires can vary in both space and time. The following classifications illustrate these potential combinations:

- **Proximate and short-term**: This combination refers to the immediate and often severe impacts experienced in the area closest to the wildfire. For example, intense heat and flames can cause immediate destruction of vegetation and habitat, with effects that are observed shortly after the fire.

- **Proximate and medium-term**: This involves impacts that occur in the immediate vicinity of the wildfire but are observed over a medium duration. These effects include soil degradation, changes in local species

(a) FCC image dated 05 August 1987 showing healthy forest in green color. The vertical line on the left is the boundary of the Yellowstone National Park.

(b) FCC image dated 02 August 1988 showing areas burnt by wildfire in red color.

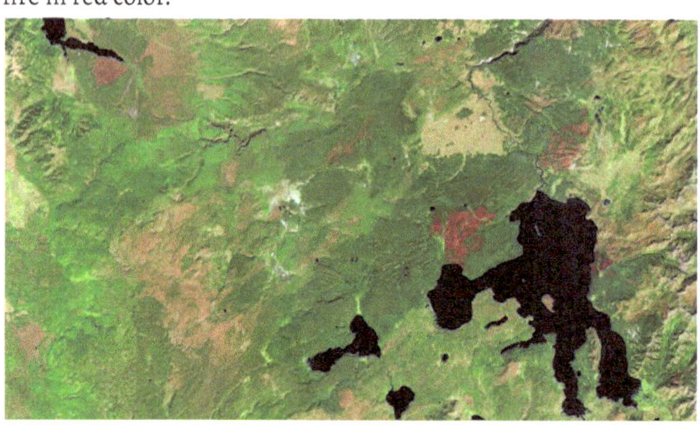

(c) FCC image dated 24 September 2011 showing that forests have still not recovered fully after more than two decades.

FIGURE 2.2: Long-term effects of the 1988 Yellowstone National Park wildfire. Images courtesy NASA's Scientific Visualization Studio.

composition, and gradual shifts in the ecosystem as the area begins to recover.

- **Proximate and long-term**: These are the enduring impacts on the area directly affected by the wildfire. They include long-lasting changes such as persistent alterations in soil chemistry, prolonged loss of biodiversity, and long-term disruption of ecological processes.
- **Intermediate and short-term**: This combination captures the effects experienced at a moderate distance from the fire in the short term. These impacts may include smoke and ash deposition, as well as changes in air quality and immediate disruptions to nearby ecosystems.
- **Intermediate and medium-term**: These effects are observed at a moderate distance from the wildfire over a medium duration. They may involve changes in vegetation patterns, gradual recovery of local ecosystems, and longer-term alterations in air quality.
- **Intermediate and long-term**: This refers to the sustained impacts at a moderate distance from the wildfire. Long-term effects may include prolonged changes in air and water quality, persistent shifts in vegetation and wildlife populations, and ongoing recovery challenges for ecosystems.
- **Distal and short-term**: This combination describes the effects experienced far from the wildfire but observed in the short term. These impacts often include the spread of smoke and pollutants, which can affect air quality and public health in distant areas.
- **Distal and medium-term**: These effects occur at greater distances from the wildfire over a medium duration. They may include continued dispersion of smoke and ash, as well as medium-term changes in atmospheric conditions and impacts on distant ecosystems.
- **Distal and long-term**: This combination involves the long-term impacts observed far from the fire. Persistent issues may include long-lasting effects on air quality, potential alterations in regional climate patterns, and ecological changes in areas far from the original wildfire.

By understanding these various combinations of spatial and temporal effects, researchers and policymakers can better assess the multifaceted impacts of wildfires. This comprehensive approach aids in the development of targeted strategies for managing and mitigating the diverse consequences of wildfires on both local and broader scales.

2. **Effects on different subject areas**: We can also analyze the effects of wildfires through various subject areas, each of which highlights different aspects of their impact. These categories include:

a. **Physical effects**: This encompasses the direct physical alterations caused by wildfires, such as changes to the landscape, soil erosion, and the destruction of physical infrastructure, mainly due to the intense heat and burning.

b. **Chemical effects**: Wildfires have significant chemical impacts, including the release of various pollutants into the atmosphere. These pollutants can include carbon monoxide, carbon dioxide, and volatile organic compounds. The combustion process also affects soil chemistry, altering nutrient availability and potentially contaminating water sources.

c. **Biological effects**: This category covers the impacts on living organisms and ecosystems. It includes the immediate harm to flora and fauna, changes in species composition, and disruptions to ecological interactions. It also involves the longer-term effects on biodiversity, habitat destruction, and the potential for species migration or extinction.

d. **Effects on society**: Wildfires have profound effects on human communities, including damage to property, loss of livelihoods, and threats to public health. They can lead to displacement of communities, economic losses, and strain on emergency services. The societal impacts also include the psychological stress experienced by affected individuals and communities.

e. **Effects on weather and climate**: Wildfires can influence local and global weather patterns and climate. The release of particulate matter and greenhouse gases can affect weather and climate, both near and far. Wildfires can also alter local weather patterns, such as increasing temperatures and altering precipitation patterns, which can have cascading effects on regional climates.

The following sections will provide a detailed examination of these effects, offering insights into the complex ways in which wildfires impact various aspects of the environment and human society.

2.2 PHYSICAL EFFECTS

Wildfires induce a variety of changes that do not necessarily involve chemical transformations. These changes are classified as physical effects and include alterations in temperature, soil structure, sediment movement through erosion, and hydrology. This section provides an in-depth analysis of these physical effects.

2.2.1 CHANGES IN TEMPERATURE

Combustion, the process of burning, is a chemical reaction that is typically exothermic, meaning that it releases energy. During a wildfire, the combustion of fuel releases energy predominantly in the form of heat and light. This results in a significant increase in temperature that is both observable and measurable.

Wildfires can achieve temperatures ranging from approximately 1,000 to 1,500 Kelvin (K). However, under extreme conditions, such as during intense wildfires, temperatures can rise to as high as 2,500 K (Dennison et al., 2006). These high temperatures can have profound impacts on the environment, including the decimation and alteration of vegetation, soil properties, and even the underlying geological structure. The heat from wildfires can cause immediate and severe changes, such as the vaporization of moisture in plants and soil, the charring of organic matter, and potential damage to infrastructure.

A significant amount of energy released by wildfires is dissipated through the processes of conduction, convection, and radiation. As a result, temperature changes can be observed at various scales — proximate, intermediate, and distal.

Temperature effects from wildfires can also be categorized into short-term, medium-term, and long-term temporal changes. As a wildfire approaches a place, the temperature rapidly escalates as more and more radiant energy starts to reach it. Once the spot starts to burn, a massive amount of energy gets released through the process of combustion. The surrounding materials start to gather and absorb this energy into their masses. After their heat capacity is reached, the temperature shoots to a maximum. Once the fuel load is depleted, and the wildfire has moved on, the temperature begins to decrease.

However, it may take several hours or even days — or years — for temperatures to return to pre-wildfire conditions, especially because the heat stored in the soil dissipates very slowly. Streams in burnt areas often exhibit elevated temperatures for several years following a large wildfire (Rhoades et al., 2011). This prolonged warming is generally accentuated by the loss of vegetation cover, which reduces shade and allows more sunlight to be absorbed, and also by the darkening of surfaces due to charring and smoke, which increases their capacity to absorb incident radiation. This elevated temperature in streams can be noticeable even at considerable distances from the wildfire, indicating distal and long-term effects. Observations have shown that the daily thermal maxima of streams in burnt areas can exceed those in unburnt areas, even when the minima are similar (Hitt, 2003).

Paradoxically, wildfires can also result in localized cooling effects (David et al., 2018). This cooling occurs through the attenuation of solar radiation by smoke (David et al., 2018; Huang et al., 2023; Kochanski et al., 2019; Stone et al., 2011). The dense smoke can act as a barrier, blocking sunlight and infrared radiation from reaching the ground. Additionally, in intermediate and distal areas, cooling can occur due to the creation of convection currents that draw in cooler air from surrounding regions

(Rosenfeld et al., 2007; Tedim et al., 2018; Zhang et al., 2019). Darkened surfaces resulting from charring can also enhance radiative cooling (Huang et al., 2023; Liu et al., 2022; Stone et al., 2011), leading to lower thermal minima, particularly during nighttime (Huang et al., 2023).

Understanding these temperature changes is crucial for assessing the broader physical impacts of wildfires, including their effects on soil stability, water retention, and overall landscape morphology.

2.2.2 CHANGES IN SOIL

Wildfires can significantly impact soils in both short and long terms.

Short-term effects: During and after a wildfire, the soil undergoes several physical and chemical changes. The intense heat from the fire can alter the physicochemical properties of the soil and its constituents. This includes changes in soil texture[1], structure[2], and chemical composition. Extreme temperatures can cause soil particles to fuse together, a process known as soil "baking," which can lead to the formation of a hardened layer on the soil surface. This layer can affect water infiltration and seed germination.

Medium- to long-term effects: Over time, wildfires can disrupt the environmental processes that influence soil properties. These include changes in the input, storage, and decomposition of both organic and inorganic soil constituents (Pellegrini et al., 2022). For example, the loss of vegetation cover following a wildfire reduces the amount of organic material added to the soil. This can lead to a decrease in soil fertility and changes in humus content and nutrient cycling. Additionally, the altered soil structure can affect water retention and erosion rates.

Soil is generally a poor conductor of heat, which means that the high temperatures generated by wildfires typically do not penetrate deeply into the soil (Ice et al., 2004). As a result, while the surface layers of soil can experience intense heating, deeper soil layers remain relatively unaffected. However, the intense baking of the surface soil can lead to several notable changes. The surface layer may become more resistant to erosion but also less permeable to water, which can exacerbate runoff and ultimately contribute to increased erosion in the post-fire environment.

A prominent change that occurs in the soil as a result of wildfires is the alteration of its structure (Chief et al., 2012; Šimanský, 2015). This refers to the way that individual particles of sand, silt, and clay are organized and assembled within the soil

[1]Soil texture refers to the proportion of different-sized particles in the soil, including sand, silt, and clay. It affects water retention, drainage, nutrient availability, and overall soil health.

[2]Soil structure refers to the arrangement and organization of soil particles into clumps or aggregates. It influences water movement, root growth, and air circulation in the soil, affecting overall soil health and fertility.

matrix. The combustion of organic matter during a wildfire leads to a significant reduction in the total organic content of the soil (Certini et al., 2011; Fernández et al., 1997).

The process of burning converts organic matter into what is known as pyromorphic[3] humus. This form of humus has notably weaker colloidal properties compared to pre-fire organic matter (González-Pérez et al., 2004). Fires also cause a sharp decrease in the concentration of lignin, a complex organic polymer, while increasing the levels of unspecific aromatics and polycyclic aromatic hydrocarbons (Jiménez-Morillo et al., 2020; Merino et al., 2018). These compounds affect soil health and its capacity to support plant growth.

In addition, the mineralized charred remains, often referred to as biochar, become incorporated into the soil (Fernández et al., 1997; Knicker, 2007). Biochar can influence soil properties by altering its structure, enhancing water retention, and potentially ameliorating soil fertility.

Wildfires have a profound impact on soil microorganisms and other living organisms in the upper layers of soil (Borgogni et al., 2019; Köster et al., 2021). These effects include both the immediate loss of these organisms and changes in their populations following the fire (Acea and Carballas, 1996; Ginzburg and Steinberger, 2012; Rodríguez et al., 2017). Consequently, these changes can be observed in the short term, as well as medium and longer terms.

The death of microorganisms and other soil organisms reduces the decomposition rate of soil organic matter (Visser and Parkinson, 1999). Thus, any organic matter in the soil may persist for extended periods (Holden et al., 2013). This persistence can further alter soil nutrient dynamics and affect ecosystem recovery.

Additionally, wildfires significantly decrease the amount of plant matter available in the area (Bartels et al., 2016; Cuevas-Gonzalez et al., 2009). Fire destroys plants, seeds, and root systems through intense heat and also creates an environment that is less conducive to plant growth. Reduction in plant matter leads to a decreased influx of organic material into the soil.

As a result of these processes, the net quantity of organic matter in the soil can vary. It may increase, decrease, or remain constant over time, depending on several factors (Kelly et al., 2021; Pellegrini et al., 2022). The final outcome is influenced by the initial soil conditions, including its structure and the availability of water, which play crucial roles in determining how the soil recovers and how organic matter dynamics evolve after the fire episode (Aedo and Bonilla, 2021).

Understanding these effects is essential for managing soil health and promoting effective ecosystem restoration following wildfires.

---

[3]The word root for 'pyromorphic' is 'pyro-,' which comes from the Greek word 'pur,' meaning 'fire.' This is combined with 'morphic,' which comes from the Greek word 'morphe,' meaning 'form' or 'shape.' Thus, 'pyromorphic' refers to something that takes on a shape or form related to fire.

2.2.3 CHANGES IN HYDROLOGY

The intense heat generated by wildfires can lead to the formation of highly hydrophobic, non-wettable soil (Chen et al., 2020b; Debano and Krammes, 1966; Varela et al., 2005; Woods et al., 2007). This increase in hydrophobicity significantly reduces the soil's ability to absorb water (Jiménez-Morillo et al., 2017), resulting in complex changes to the hydrology of the affected region (Doerr et al., 2006). These changes include increased surface runoff, elevated flood risks (Moody et al., 2008; Prosser and Williams, 1998), and alterations to geomorphology (Moody and Martin, 2001; Shakesby and Doerr, 2006), erosion patterns (Huffman et al., 2001; Shakesby, 2011), and ecological dynamics (Kinnaird and O'Brien, 1998).

Wildfires typically reduce soil water retention capabilities (Ebel, 2012) due to structural changes and increased hydrophobicity. Reduction in water absorption decreases the moisture available to plants, adversely affecting their growth and having long-term implications for future wildfire occurrences (Hou et al., 2020; Jensen et al., 2018; Krueger et al., 2022).

Moreover, changes in soil properties and increased runoff lead to the mobilization of large quantities of solutes and sediments into nearby water bodies and reservoirs (Moody and Martin, 2004; Warrick and Rubin, 2007). The deposition of these sediments can alter hydrological connectivity (Ortíz-Rodríguez et al., 2019; Wu et al., 2021) and disrupt streamflow patterns (Gao et al., 2012; Guo et al., 2023; Khaledi et al., 2022), as large sediment loads can obstruct stream channels. Additionally, sediments can affect groundwater recharge rates by clogging soil pores (Rey et al., 2023) and reduce water storage capacity of watersheds (Hallema et al., 2018; Nyman et al., 2014). This occurs as sediments fill low-lying areas and impede aquifer recharge.

These changes collectively contribute to a reduction in water quality and availability in immediate as well as distant areas surrounding the wildfire (Emmerton et al., 2020; Rhoades et al., 2019; Smith et al., 2011). Understanding these hydrological impacts is crucial for managing water resources and mitigating the broader ecological consequences of wildfires.

2.3 CHEMICAL EFFECTS

In addition to physical changes, wildfires cause significant chemical alterations in the environment. These changes can degrade air and water quality and have far-reaching effects, extending well beyond the immediate area of the fire.

2.3.1 RELEASE OF HAZARDOUS SUBSTANCES

Wildfires involve the combustion of biological materials rich in carbon, hydrogen, oxygen, nitrogen, sulfur, and phosphorus. The intense heat causes these materials to undergo chemical transformations, releasing a variety of hazardous substances:

1. **Carbon compounds**: A major component of wildfire emissions is carbon dioxide (CO_2), carbon monoxide (CO), and methane (CH_4). Carbon dioxide is a prominent greenhouse gas that contributes to global warming. Excess CO_2 can also reduce oxygen levels, posing risks to respiratory health. Carbon monoxide, which binds more effectively to hemoglobin than oxygen, forms carboxyhemoglobin, reducing blood's oxygen-carrying capacity. This makes carbon monoxide an extremely toxic gas. Distribution of atmospheric carbon monoxide during Camp wildfire, together with baseline data is shown in figure 2.3a–b[4]. We can observe a distinct increase in atmospheric carbon monoxide concentration at the time of the Camp wildfire. Methane, another potent greenhouse gas, has a global warming potential 84 times greater than carbon dioxide over 20 years and 28 times greater over 100 years (Skytt et al., 2020). Methane is also flammable and can displace oxygen, leading to asphyxiation (Byard, 2011, 2023; McEwen, 2018; Sauvageau and Boghossian, 2010).

2. **Pyrogenic hydrocarbons and Non-Methane Organic Compounds (NMOCs)**: Wildfires release various pyrogenic[5] hydrocarbons and NMOCs, including methanol, formaldehyde, acetaldehyde, acetone, hydroxyacetone, acetic acid, furan, butanedione, alkanes, alkenes, alkynes, and volatile organic compounds (VOCs) such as benzene, toluene, ethylbenzene, and xylene (Liu et al., 2014, 2017; Urbanski et al., 2008; Urbanski, 2014). Many of these substances are toxic, carcinogenic (cancer-causing), and teratogenic[6], potentially causing deaths, cancers, and developmental disorders. They can be absorbed through respiration, contact, and ingestion, and are often bioaccumulated[7] in the bodies of organisms. Distribution of atmospheric formaldehyde during Camp wildfire, together

[4]At approximately 2.3 μm (Short Wave Infra Red, SWIR band), carbon monoxide (CO) displays distinct absorption bands attributed to the vibrational transitions of its molecular bonds. As sunlight traverses the atmosphere, CO molecules absorb specific wavelengths of light, leading to a reduction in the intensity of light at those wavelengths. By quantifying the amount of sunlight absorbed at those wavelengths, instruments can deduce the concentration of CO present in the atmosphere. The figure has been created with data from Sentinel-5P satellite, and carbon monoxide has been represented as vertically integrated CO column density, in the units of moles per square meter of the Earth's surface. The represents the number of moles of carbon monoxide present in a column of atmosphere with a ground footprint of one square meter.

[5]The word root for 'pyrogenic' is 'pyro-,' which comes from the Greek word 'pur,' meaning 'fire.' The suffix '-genic' means 'producing' or 'causing.' So, 'pyrogenic' refers to something that is produced by or related to fire.

[6]'Teratogenic' comes from the Greek roots 'terato-' meaning 'monster' or 'malformation,' and '-genic,' meaning 'producing' or 'causing.' So, teratogenic refers to agents that cause malformations or developmental abnormalities in embryos or fetuses.

[7]Bioaccumulation happens when organisms accumulate toxic substances in their bodies faster than they can eliminate them.

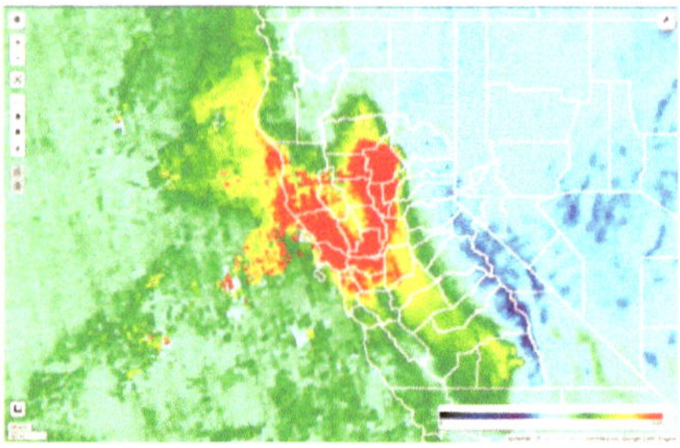

(a) Distribution of atmospheric carbon monoxide in November 2018 at the time of Camp wildfire.

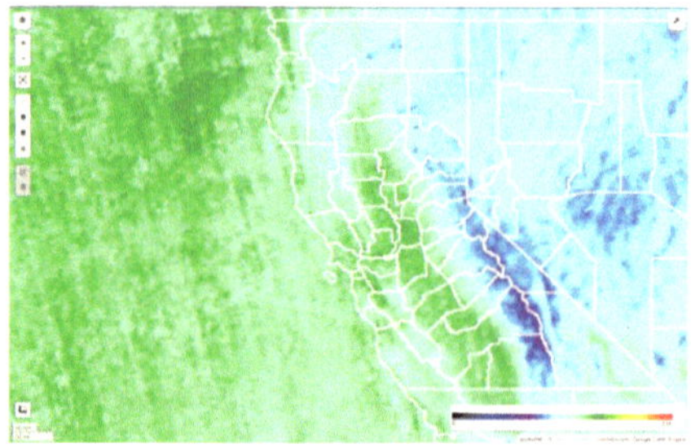

(b) Baseline distribution of atmospheric carbon monoxide in November 2019, a year after Camp wildfire, for comparison.

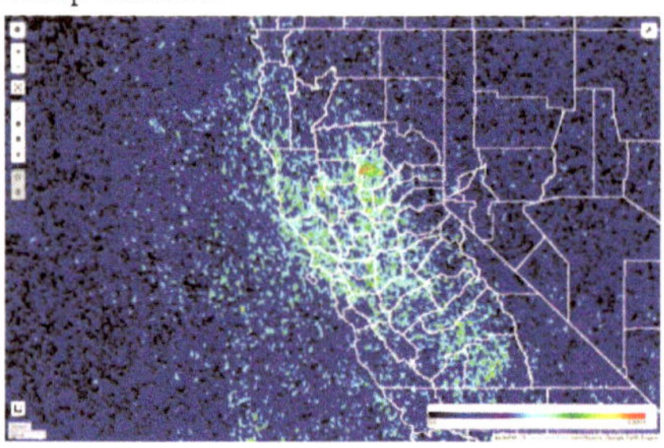

(c) Distribution of atmospheric formaldehyde in November 2018 at the time of Camp wildfire.

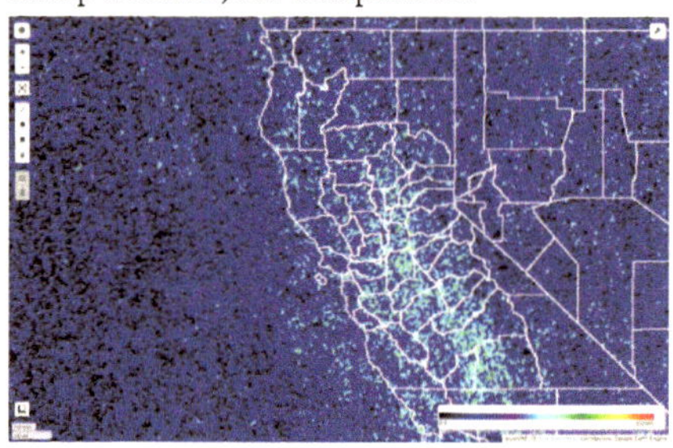

(d) Baseline distribution of atmospheric formaldehyde in November 2019, a year after Camp wildfire, for comparison.

FIGURE 2.3: Distribution of some carbon compounds around Camp wildfire as discerned through Sentinel-5P satellite. Data courtesy European Space Agency, Open Street Map, and Open Topo Map. Visualized on geemap using Google Earth Engine.

with baseline data, are shown in figure 2.3c–d[8]. We can observe a distinct increase in the concentration of atmospheric formaldehyde at the time of Camp wildfire, much above the general level present due to other sources of pollution.

3. **Halogen-containing chemicals**: Wildfires produce halogenated compounds such as hydrochloric acid, methyl chloride, methyl bromide,

[8]The measurement of formaldehyde (HCHO) from space is facilitated by sophisticated remote sensing techniques that exploit solar ultraviolet (UV) backscatter within the wavelength range of 325 to 360 nm. Instruments such as the Tropospheric Monitoring Instrument (TROPOMI) utilize the distinct absorption properties of formaldehyde in this spectral region to detect its atmospheric presence. When sunlight interacts with atmospheric HCHO, the molecules scatter and reflect light back to the instrument, thereby enabling the identification and quantification of formaldehyde concentrations. The figure has been created with data from Sentinel-5P satellite, and formaldehyde has been represented as vertically integrated formaldehyde column density, in the units of moles per square meter of the Earth's surface. This represents the number of moles of formaldehyde present in a column of atmosphere with a ground footprint of one square meter.

methyl iodide, polychlorinated dibenzo-p-dioxins, polychlorinated dibenzofurans, and polychlorinated biphenyls (Urbanski et al., 2008; Young et al., 2004; Zhang et al., 2017). These compounds are respiratory and skin irritants, and many are known carcinogens and teratogens. Dioxins, in particular, are persistent organic pollutants (POPs) that are very slow to degrade and have long-term toxic effects on reproductive, developmental, immune, and endocrine systems.

4. **Nitrogen compounds**: Wildfires release nitrogen compounds, including nitrogen oxides (NO_x), nitrous acid, ammonia (NH_3), acetonitrile (CH_3CN), cyanohydrins, hydrogen cyanide (HCN), cyanogen ($(CN)_2$), and other cyanides from plants containing cyanogenic glycosides (Andreae and Merlet, 2001; Andreae, 2019; Flematti et al., 2011; Urbanski et al., 2008). Nitrogen oxides contribute to photochemical smog, acid rain, and tropospheric ozone formation (Dickerson et al., 1997; Pitts Jr and Finlayson, 1975). These compounds convert into secondary pollutants[9] such as nitrous and nitric acids, dinitrogen pentoxide, peroxyacetyl nitrate (PAN), and peroxynitric acid, which are respiratory and eye irritants (Harrison, 1986; Sillman, 2003). Acidic compounds in rainfall can damage trees, disrupt ecosystems, and alter biogeochemical cycles (Gorham, 1976; Irwin and Williams, 1988; Johnson et al., 1982; Likens et al., 1996). Cyanides, including hydrogen cyanide, are highly toxic and can inhibit cytochrome c oxidase, disrupting cellular respiration (Antonini et al., 1971; Leavesley et al., 2008). Compounds like acetonitrile can be metabolized into hydrogen cyanide within organisms, further exacerbating toxicity (Freeman and Hayes, 1987; Organization et al., 1993). Distribution of atmospheric nitrogen dioxide during Camp wildfire, together with baseline data, is shown in figure 2.4[10]. We can observe a distinct increase in atmospheric nitrogen dioxide concentration at the time of Camp wildfire. This is in addition to nitrogen dioxide present due to various human activities[11]. Urban environments

[9]A secondary pollutant is a harmful substance that is not emitted directly into the air but forms in the atmosphere through chemical reactions involving primary pollutants.

[10]Advanced remote sensing techniques using solar blue backscatter between 400 to 470 nanometers measure nitrogen dioxide from space. Instruments like TROPOMI (Tropospheric Monitoring Instrument) detect nitrogen dioxide's unique absorption in this spectral region by measuring the light scattered and reflected back to the instrument. The figure has been created with data from Sentinel-5P satellite, and nitrogen dioxide has been represented as tropospheric vertically integrated nitrogen dioxide column density, in the units of moles per square meter of the Earth's surface. This represents the number of moles of nitrogen dioxide present in a column of the troposphere, the lowest layer of the atmosphere, with a ground footprint of one square meter.

[11]Nitrogen dioxide (NO_x) is primarily generated from anthropogenic sources, particularly motor vehicle emissions. During the combustion process in internal combustion engines, high temperatures and pressures facilitate the reaction between nitrogen (N_2) and oxygen (O_2) in the air, resulting in the formation of nitrogen oxides (NO_x), including NO_2. Industrial activities, such as power generation and manufacturing, further contribute to ambient NO_2 levels through the combustion of fossil fuels. Additionally, residential heating systems, especially those utilizing gas, are notable sources of nitrogen dioxide.

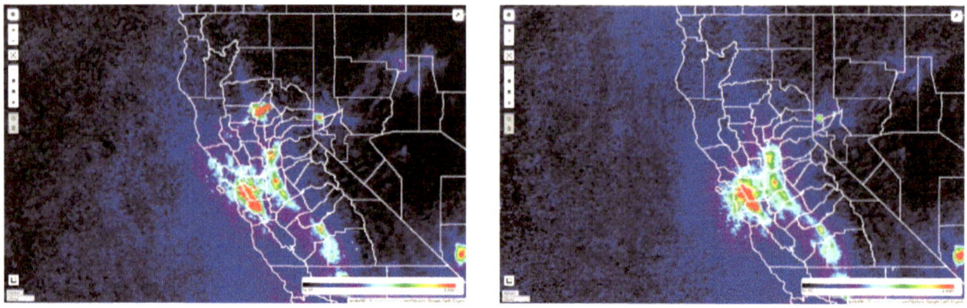

(a) Distribution of atmospheric nitrogen dioxide in November 2018 at the time of Camp wildfire.

(b) Baseline distribution of atmospheric nitrogen dioxide in November 2019, a year after Camp wildfire, for comparison.

FIGURE 2.4: Distribution of atmospheric nitrogen dioxide around Camp wildfire as discerned through Sentinel-5P satellite. Data courtesy European Space Agency and Open Street Map. Visualized on geemap using Google Earth Engine.

typically exhibit elevated concentrations of NO_2, attributable to the high density of vehicular traffic and industrial operations. In the figure, we observe a new spot of nitrogen dioxide concentration near the site of Camp wildfire, that is not present in the baseline data.

5. **Sulfur compounds**: Sulfur-containing compounds released during wildfires include hydrogen sulfide, sulfur oxides (SO_x), carbonyl sulfide, dimethyl sulfide, and various sulfates (Flematti et al., 2011; Liu et al., 2014, 2017; Urbanski et al., 2008; Urbanski, 2014). Hydrogen sulfide is a corrosive, flammable, explosive, and poisonous gas with toxicity comparable to carbon monoxide (Rubright et al., 2017). Sulfur oxides react with water vapor to form sulfurous and sulfuric acids, leading to acid rain (Irwin and Williams, 1988). Carbonyl sulfide is an irritant and can cause respiratory paralysis in high concentrations (Bartholomaeus and Haritos, 2005; Peyton et al., 1978), while dimethyl sulfide is highly flammable and an irritant to the eyes and skin (Authority, 2020). Sulfates contribute to PM2.5 particulate matter, impacting air quality and respiratory health (Schlesinger, 2007).

6. **Inorganic particulate matter**: Inorganic emissions from wildfires include particulate matter and ash containing calcium, magnesium, potassium, silicon, phosphorus, sodium, sulfur, aluminum, iron, manganese, and zinc. At temperatures around 500°C, silica and carbonate compounds, such as calcium carbonate, magnesium carbonate, and potassium carbonate, are prevalent in ash. Above 500°C, these carbonates decompose into oxides (Bodí et al., 2014). These inorganic compounds alter the chemical properties of soil and water, affecting pH, disrupting ecosystems, and in-

terfering with nutrient cycles (Grantz et al., 2003; Rai, 2016). Particulates can also lead to health issues like asthma, lung cancer, silicosis, developmental disorders, neurodegenerative disorders, and premature death (Schlesinger, 2007).

7. **Other hazardous substances**: Plants can accumulate hazardous substances, including heavy metals (e.g., mercury, nickel, cadmium, arsenic, lead, and chromium), pesticides, herbicides, and polychlorinated biphenyls (PCBs) in their bodies (Ali et al., 2013; Jabeen et al., 2009; Muthusaravanan et al., 2018). When these plants burn, these hazardous substances are released back into the environment. Radionuclides[12] can also be released. For instance, during the 2020 wildfires in the Chernobyl Exclusion Zone, significant amounts of radionuclides were emitted: approximately 341 GBq[13] of cesium-137 (^{137}Cs), 51 GBq of strontium-90 (^{90}Sr), 2 GBq of plutonium-238 (^{238}Pu), 33 MBq[14] of plutonium-239 (^{239}Pu), 66 MBq of plutonium-240 (^{240}Pu), and 504 MBq of americium-241 (^{241}Am) (Evangeliou and Eckhardt, 2020; Igarashi et al., 2020; Masson et al., 2021). In a similar way, cesium-137 deposited in Sweden's forests following the Chernobyl accident was also released through wildfires (Martinsson et al., 2021). These radionuclides are particularly harmful to developing animals and humans, increasing the risk of cancer, immune deficiency, and radiation sickness (Kamiya et al., 2015; Lindell, 1987).

2.3.2 DEGRADATION OF AIR AND WATER QUALITY

Wildfires significantly impair air and water quality through the emission of various pollutants. Among the byproducts of combustion are volatile organic compounds (VOCs) such as benzene, formaldehyde, and acetaldehyde, which are classified as hazardous air pollutants (HAPs). These HAPs are associated with numerous health issues, including asthma, cancer, and chronic obstructive pulmonary disease (COPD) (Choi et al., 2018; Leikauf, 2002; Loh et al., 2007). The presence of these pollutants, along with aerosols and particulates, diminishes air and water quality, contributing to degradation of various habitats (Mehaffey et al., 2009; Zvereva and Kozlov, 2010).

Following wildfires, stream temperatures often rise significantly (Rhoades et al., 2011), which can profoundly affect various water characteristics. Increased temperatures influence the solubility of salts — making most salts more soluble in wa-

[12] Radionuclides are unstable atoms that emit radiation as they decay into more stable forms. They can occur naturally or be produced artificially. Some common examples include cesium-137 and uranium-238.

[13] GBq stands for gigabecquerel, a unit of radioactivity in the International System of Units (SI). It represents one billion (10^9) disintegrations per second.

[14] MBq stands for megabecquerel, a unit of radioactivity equal to one million (10^6) disintegrations per second.

ter, and decreasing the dissolution of oxygen and other gases in the water. These changes can lead to chemical toxicity in aquatic environments (Hitt, 2003), irrespective of whether or not the increase in temperature, by itself, exceeds the temperature tolerance limits of different species (Isaak et al., 2010; Meador and Carlisle, 2007).

The degradation of water quality is also compounded by the influx of ash, sediments, and nutrients from wildfires (Smith et al., 2011). This additional load further disrupts the aquatic ecosystem by altering the chemical composition of water, affecting both its physical properties and its capacity to support diverse forms of life. Consequently, the combined effects of elevated temperatures and wildfire-induced contaminants contribute to significant declines in water quality and ecosystem health.

The degradation of air and water quality is further exacerbated by the generation of new and potentially hazardous compounds. Wildfires significantly alter atmospheric chemistry by releasing a variety of reactive substances. These include free radicals (Leonard et al., 2007; Sigmund et al., 2021), reactive nitrogen species (Benedict et al., 2017; Lindaas et al., 2021), hydroxyl radicals (OH) (Fang et al., 2023), ozone (Jaffe and Wigder, 2012), and reactive oxygen species (Fang et al., 2023). These reactive compounds contribute to the formation of secondary pollutants and influence atmospheric reactivity.

The transformation of organic chemicals during wildfires also produces several new compounds. These include secondary organic aerosols[15] (SOAs) (Liang et al., 2022; Tomaz et al., 2018), which can affect air quality and climate. Additionally, compounds such as peroxyacetyl nitrate (PAN) (Peng et al., 2021), polycyclic aromatic hydrocarbons (PAHs) (Kieta et al., 2022), dioxins (Ward and Lincoln, 2006), furans (Newland et al., 2022), and other persistent organic pollutants (POPs) (Gong and Wang, 2021) are also released. These pollutants are known for their environmental persistence and potential to cause long-term harm to ecosystems and human health (Alharbi et al., 2018; Qing Li et al., 2006).

2.4 BIOLOGICAL EFFECTS

Wildfires significantly impact ecosystems and biological communities. They can destroy habitats, leading to immediate loss of biodiversity and disruption of food webs. Many species, particularly those that are non-adapted to fire, suffer from population declines or displacement. On the other hand, wildfires can also play a crucial role in certain ecosystems by promoting the regeneration of fire-adapted plants, clearing dead material, and returning nutrients to the soil. This cyclical process can enhance biodiversity in the long term, as new growth attracts a variety of species. However, the increasing frequency and intensity of wildfires, often exacerbated by climate change and human activity, are pushing biological communities

[15]Secondary organic aerosols (SOAs) are tiny particles formed in the atmosphere through the chemical reactions of volatile organic compounds (VOCs). When VOCs react with sunlight and other atmospheric components, they can create SOAs. These particles play a significant role in air quality, climate change, and human health, and they can affect cloud formation and visibility.

to changes at a pace much faster than their pace of adaptation. This poses serious challenges to ecosystem resilience and biodiversity conservation.

2.4.1 CHANGES IN HABITAT

Wildfires induce profound alterations in habitats across various scales, impacting terrestrial, aquatic, and aerial environments. The immediate and long-term effects on these habitats can be significant and multifaceted.

Wildfires lead to the degradation and destruction of terrestrial habitats as trees and plants are consumed by flames (Bosso et al., 2018; McGarigal et al., 2005). This destruction reduces the availability of critical resources such as food (Apfelbaum and Haney, 1981; Nkwabi et al., 2011), shelter, and nesting sites for wildlife (Brooker and Rowley, 1991; Peterson et al., 2023; Stillman et al., 2019), as depicted in figure 2.5. The loss of vegetation can destabilize soil, increase erosion, and contribute to the loss of habitat complexity, so crucial for supporting diverse species.

Wildfires also adversely affect aquatic habitats through several mechanisms. Elevated temperatures from wildfires can raise water temperatures, negatively impacting aquatic organisms (Schindler, 2017). Additionally, hazardous chemicals from wildfires can be transported into water bodies, further compromising water quality (Scordo et al., 2021). Runoff from burned areas can introduce excessive nutrients into aquatic systems, potentially leading to chemical toxicity and eutrophication (Vashukevich et al., 2023). This process promotes the rapid growth of aquatic plants, which can obstruct animal movement and deplete oxygen levels as plant biomass decomposes. The resulting anoxic conditions can lead to widespread fish kills and disrupt aquatic ecosystems.

The impact on aerial habitats is also significant. Smoke, heat, and airborne chemicals from wildfires can impair the health of birds and other aerial species (Stuart-Smith et al., 2002). Smoke can reduce air quality and visibility, affecting foraging and migratory behavior of organisms. High temperatures and presence of toxic substances in the atmosphere can also lead to respiratory issues and other health problems in birds.

Wildfires profoundly disrupt soil microbial communities and cause widespread destruction that alters the nutrient cycling dynamics within an ecosystem (Eckdahl et al., 2023; Elliott et al., 2013). The intense heat and combustion associated with wildfires can kill beneficial microorganisms crucial for decomposing organic matter and recycling nutrients. This disruption can lead to changes in the availability of nutrients, affecting plant growth and overall ecosystem productivity.

In addition to disrupting nutrient cycling, wildfires can also modify the microclimate[16] of the affected area (Brown et al., 2014; Wolf et al., 2021). Loss of vegetation

[16]A microclimate is a small, localized climate that differs from the surrounding area. It is often influenced by factors like topography, vegetation, and human activity. Examples include shaded areas under

(a) The pygmy hog is an endangered species, and the only species in the genus *Porcula*. The population, once spread throughout the foothills of the Himalayas, is currently restricted only to the state of Assam in India.

(b) Wildfires in the grasslands of Assam threaten pygmy hogs in their last remaining habitat, pushing them towards extinction.

(c) Veld fires in South Africa create vast desolate landscapes and conditions of food insecurity for species like baboons.

(d) Wildfires in Cape Town reduce the availability of food and cover for bonteboks and other species.

FIGURE 2.5: Some effects of wildfires on wildlife habitats.

and changes in soil properties can alter temperature and moisture levels, further influencing the local climate and affecting the survival of plant and animal species. Such changes can have cascading effects on the ecosystem, influencing everything from species composition to habitat structure (Awadhiya, 2021).

Long-term impacts of these disruptions can lead to significant habitat changes and may even reset the process of ecological succession (Dawe et al., 2021). Ecological succession refers to the gradual process of change and development in an ecosystem following a disturbance. Wildfires, especially when they occur frequently, can reset this process, leading to the formation of a new, often less diverse community known as a disclimax community (Awadhiya, 2021), which is characterized by reduced heterogeneity and decreased biological diversity compared to pre-fire condi-

trees, urban heat islands, or regions near bodies of water. Microclimates can significantly affect local flora and fauna.

tions (Zhang et al., 2021; Pastro et al., 2011). This shift can alter the structure and function of the ecosystem, leading to long-term changes in habitat quality and the ability of the area to support a diverse range of species.

Wildfires also have significant impacts on soil seed banks and wildlife, leading to profound ecological changes. Many soil seed banks, which are crucial for the regeneration of plant communities, are destroyed by the intense heat of wildfires (Maia et al., 2012). The loss of these seed banks can delay or inhibit the recovery of vegetation, affecting the overall resilience of the ecosystem.

In addition to affecting plant life, wildfires also displace numerous animal species and birds from their natural habitats. The destruction of native species and their habitats creates opportunities for invasive species to establish themselves (Rew and Johnson, 2010). Invasive species can in turn alter the fire regime of the area, often by making forests more susceptible to future wildfires (Faccenda and Daehler, 2022). This shift can create a cycle where frequent fires facilitate the spread of more fire-prone (and even fire-dependent) invasive species, further exacerbating the problem.

Larger animals requiring extensive home ranges are particularly vulnerable to habitat loss caused by wildfires. Species of top predators (e.g., tigers) and ecosystem engineers (e.g., elephants) are especially severely impacted since they often have large home range requirements. Local extinction of these animals can lead to trophic cascades, where the absence of a key species disrupts the balance of the entire ecosystem, resulting in long-term ecological changes (Awadhiya, 2021; Banks et al., 2011). For example, loss of top predators can lead to an overabundance of prey species, which in turn affects vegetation and other aspects of the ecosystem.

Conversely, smaller, r-selected organisms, such as rodents and insects, may thrive in the vacant niches left by wildfires. These organisms reproduce rapidly and can proliferate in the disturbed environment. While their increased numbers can contribute to ecological succession, they also have several potential downsides. Many of these smaller organisms are carriers of diseases, and their proliferation can impact food security and public health in the region. Their presence may also contribute to the spread of zoonotic diseases and affect the overall epidemiology of the area.

Destruction of critical habitats and habitat connectivity corridors caused by wildfires has particularly severe consequences for endangered species (Butcher, 2019; Khosravi et al., 2022; Tracey et al., 2018). These species often rely on specific and interconnected habitats to survive and thrive. The loss of these habitats can significantly impact their populations and ability to persist and survive.

Wildfires often lead to the death of unique and rare species (Ager et al., 2007), as well as to a reduction in genetic diversity. Loss of genetic diversity is crucial because it affects the species' ability to adapt to changing conditions and recover from environmental stressors. Additionally, wildfires can disrupt wildlife reproduction and population recruitment, further exacerbating the threats faced by endangered species (Awadhiya, 2021; Lam et al., 2020; Potvin et al., 2017).

The combined effects of habitat destruction, loss of genetic diversity, and disruptions to reproductive processes can accelerate extinction vortices, where populations decline rapidly and become increasingly vulnerable to extinction (Bond and Bradley, 2003; Lindenmayer and Possingham, 1995; Santelices et al., 2022). As endangered species struggle to survive in the face of these compounded threats, they may be pushed towards complete extinction.

Extinction of species has further lasting impacts on the ecosystem, potentially altering the ecological balance and functioning of the area permanently. Loss of key species can disrupt ecological processes, food webs, and habitat structures, leading to long-term changes in the ecosystem's composition and function.

The impact is particularly severe for species and biological communities that are already under stress. Habitat degradation — driven by the introduction of pollutants and physical changes induced by wildfires — often precedes habitat loss. Such degradation can have significant consequences for biodiversity, potentially contributing to the extinction of species over intermediate and long-term periods.

In this way, wildfires pose a critical threat to endangered species by destroying their habitats, reducing genetic diversity, and disrupting reproductive processes, pushing species toward extinction and causing permanent changes to the ecosystems they inhabit. Addressing these challenges requires targeted conservation efforts and strategies to mitigate the impacts of wildfires and support the recovery of affected species and ecosystems.

2.4.2 DEATH AND DISPLACEMENT OF ORGANISMS

Wildfires result in the death of numerous plants and animals, either directly through burning or indirectly through subsequent exposure to smoke and hazardous chemicals. The impacts extend beyond immediate mortality, as many organisms suffer lingering deaths due to habitat destruction, environmental stressors, and loss of food sources. The destruction of vegetation not only leads to an immediate loss of carbon sequestration capacity but also consequently reduces the availability of food for herbivores, which in turn affects the entire food chain.

Wildfires significantly alter wildlife behavior and migration patterns. Destruction of habitats and changes in community structures are known to modify wildlife behavior (Sanderfoot et al., 2021) and migration patterns (Lewis et al., 2022; Overton et al., 2022). Simplification of ecological communities due to species loss and disruption of species interactions reduces the resistance and resilience of the ecosystem to future disturbances (Brehme et al., 2011; Davies et al., 2012; McWethy et al., 2019). This process not only exacerbates the long-term ecological impacts of wildfires but also accelerates the risk of extinction for various organisms.

The combined effects of habitat destruction, altered migration patterns, and decreased ecological resilience create a cascading series of consequences that extend well beyond the immediate aftermath of a wildfire. These changes contribute to a gradual erosion of biodiversity and can push species closer to extinction, further destabilizing ecosystems and reducing their capacity to recover from future disturbances.

2.5 EFFECTS ON SOCIETY

Wildfires impose significant human, social, and economic costs (Bayham et al., 2022; Paveglio et al., 2015; Thomas et al., 2017). They frequently result in the loss of life and property and cause a range of short- and long-term health issues. These health impacts include heat-related illnesses, dehydration, burns, and respiratory problems from inhaling smoke and hazardous chemicals. The movement of smoke and air pollutants can degrade indoor air quality over vast areas, representing one of the more insidious effects of wildfires (Liang et al., 2021).

Additionally, the contamination of water bodies poses a threat to drinking water supplies for many communities (Robinne et al., 2018). Contaminants such as ash, sediments, nutrients, and chemical compounds can also impair water treatment processes, reducing their effectiveness and leading to frequent equipment failures (Chen et al., 2020a; Hohner et al., 2019).

The economic burden of wildfires includes costs related to lost workdays due to health issues, reduced visibility, and pollution, as well as the financial strain of repairing infrastructure and allocating resources for long-term recovery, restoration, and rehabilitation efforts (Kim and Jakus, 2019; McConnell et al., 2021).

Wildfires also increase the risk of secondary hazards[17] such as soil erosion, flooding, landslides, and debris flows, and these together with the deposition of smoke and soot lead to the loss of ecosystem services and recreational opportunities (Gellman et al., 2022). There are also legacy effects[18] impacting infrastructure vulnerability and negatively influencing agriculture and forestry (Fraser et al., 2022).

The displacement of people and communities disrupts social networks, community cohesion, and cultural traditions. Indigenous communities and forest-dwellers, in particular, face profound losses in cultural, religious, and historical resources such as traditional lands, sacred sites, sacred groves, ethno-medicinal plants, and indigenous food systems. These losses have significant historical, spiritual, and cultural ramifications, leading to a loss of cultural landscape and identity, and can also result in long-term psycho-social impacts (Eisenman et al., 2015; Mihalus et al., 2023).

---

[17] A secondary hazard is a risk or danger that arises as a consequence of an initial event or primary hazard, often further exacerbating the situation. For instance, after an earthquake (the primary hazard), secondary hazards can include landslides, tsunamis, or fires triggered by damaged infrastructure. These hazards can complicate rescue efforts and increase overall damage.

[18] Legacy effect is the impact of past actions, decisions, or conditions on present and future situations, often leading to enduring consequences in systems, behaviors, or environments.

On a global scale, the world community bears the cost of pollutants and the social cost of carbon — a measure of the marginal damage from the emission of one additional tonne of carbon dioxide. This cost reflects the broader impacts of global warming and climate change, highlighting the extensive and interconnected nature of wildfire impacts (Sweeney et al., 2023).

2.6 EFFECTS ON WEATHER AND CLIMATE

Wildfires exert substantial impacts on both short-term weather patterns and long-term climate dynamics. The intense heat generated by wildfires increases atmospheric instability[19], initiating turbulence and convection currents. As hot air rises, it cools and condenses, leading to the formation of pyrocumulus clouds — fluffy, cotton-like clouds that originate from fire (from the Greek word root *pur*, meaning fire) (Lareau and Clements, 2016). Additionally, aerosols and particulate matter released during wildfires serve as nucleation sites for cloud formation, further contributing to changes in the local weather.

The smoke released from wildfires also alters air chemistry and composition, creating regional haze (Budisulistiorini et al., 2018). Haze can be quantified in terms of aerosol optical depth, a measure of the extinction of the solar beam by haze-causing particles in the atmosphere, including dust, smoke, and aerosol pollution. Figure 2.6 shows the increase in aerosol optical depth during the Camp wildfire, to levels much higher than those at other baseline times[20]. This haze, along with modified cloud properties — such as changes in droplet size, lifetime, and reflectivity — further affects local microclimates and weather patterns (Duff et al., 2014). The combined effect of clouds and haze can reduce the amount of sunshine or insolation reaching the Earth's surface.

Following wildfires, hot and dry conditions can modify air masses[21] and alter the atmospheric boundary layer[22], which is the lowest part of the troposphere. This alteration changes wind patterns and affects the atmospheric transport of pollutants

[19]Atmospheric instability refers to a condition in the atmosphere where air parcels can rise freely, leading to the development of clouds and precipitation. It occurs when warm, less dense air is situated below cooler, denser air, causing the warm air to rise. This instability can result in turbulent weather patterns, including thunderstorms and severe storms.

[20]The figure represents aerosol optical depth over land retrieved in the MODIS Blue band (0.47 μm) and the MODIS Green band (0.55 μm).

[21]An air mass is a large body of air that has uniform temperature and humidity characteristics throughout. It forms when air remains over a particular region for an extended period, acquiring the properties of that area. Air masses are typically classified based on their source region — such as tropical, polar, maritime, or continental — and significantly influence weather patterns.

[22]The atmospheric boundary layer is the lowest part of the atmosphere, typically extending from the Earth's surface up to about 1 to 2 kilometers (0.6 to 1.2 miles) high. It is characterized by significant turbulence and mixing due to surface heating, friction, and other factors. This layer plays a crucial role in weather and climate, as it directly interacts with the Earth's surface, influencing temperature, moisture, and wind patterns.

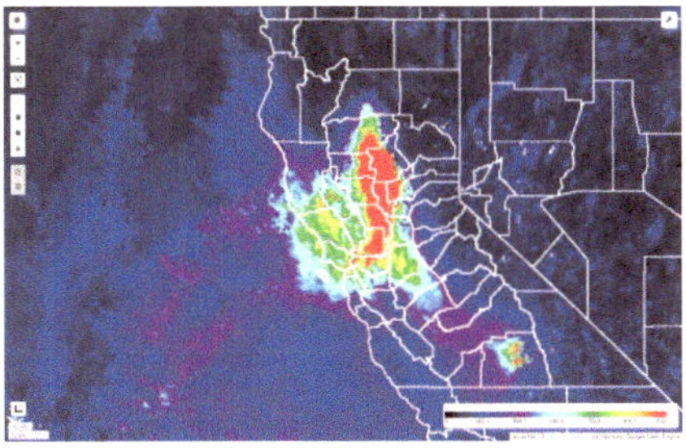

(a) Distribution of atmospheric haze in November 2018 at the time of Camp wildfire, measured in blue band (470 nm).

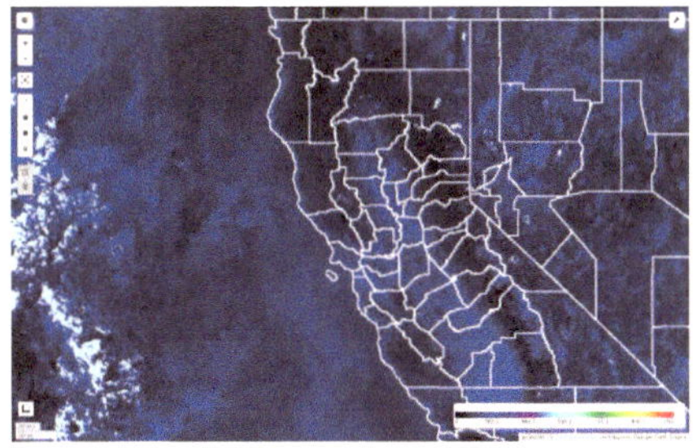

(b) Baseline distribution of atmospheric haze in November 2019, a year after Camp wildfire, for comparison, as measured in blue band (470 nm).

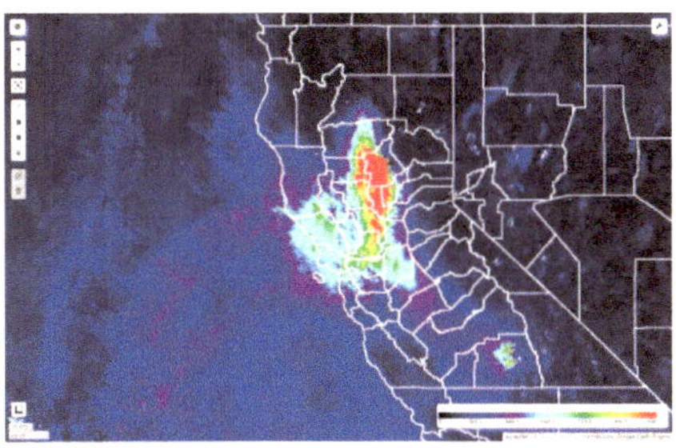

(c) Distribution of atmospheric haze in November 2018 at the time of Camp wildfire, measured in green band (550 nm).

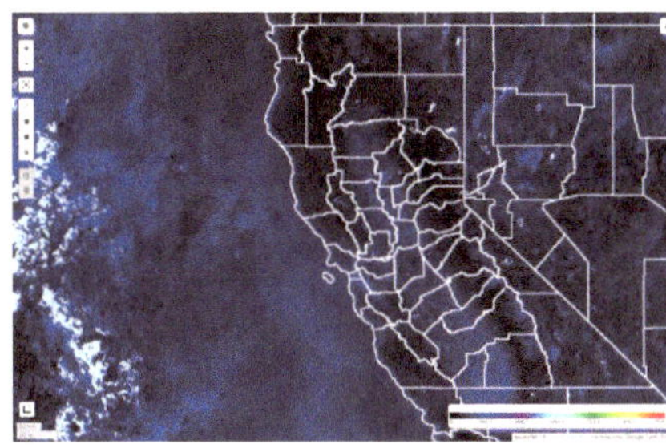

(d) Baseline distribution of atmospheric haze in November 2019, a year after Camp wildfire, for comparison, as measured in green band (550 nm).

FIGURE 2.6: Distribution of haze around Camp wildfire as discerned through MODIS aerosol optical depth measurements on Terra and Aqua satellites. Data courtesy NASA and Open Street Map. Visualized on geemap using Google Earth Engine.

in the short term, while also influencing precipitation patterns and the hydrological cycle in the long term (Shakesby and Doerr, 2006).

Over time, reduced vegetation cover and soot deposition lead to changes in albedo — the fraction of light reflected by a surface — further affecting the climate system (Gatebe et al., 2014). Soot deposited on snow modifies snowpack dynamics[23] and accumulation patterns, impacting the cryosphere[24] (Awadhiya, 2021; Aubry-Wake

---

[23]Snowpack dynamics refers to the processes and changes occurring within a layer of accumulated snow, including its formation, compaction, melting, and interactions with the environment.

[24]The cryosphere is the portion of the Earth's system that includes all forms of frozen water, such as glaciers, ice caps, sea ice, and permafrost. It plays a vital role in the global climate system by influencing sea levels, weather patterns, and the Earth's energy balance. The cryosphere is sensitive to climate change, with rising temperatures leading to melting ice and alterations in its extent and thickness.

et al., 2022).

Wildfires also release greenhouse gases and affect the carbon cycle, creating climate forcings[25] that influence atmospheric conditions. These forcings contribute to increased variability in climatic conditions. Many of the long-term climate feedbacks resulting from wildfires are positive feedbacks — effects that amplify the initial climate change — and these exacerbate extreme weather events (Halofsky et al., 2020; McKenzie et al., 2004; Richardson et al., 2022; Westerling and Bryant, 2008). For instance, the warming caused by wildfires can intensify droughts, and the resultant high temperatures and reduced moisture increase the likelihood of subsequent large wildfires, creating a vicious cycle of escalating climate impacts.

In summary, wildfires significantly alter both weather and climate, leading to immediate changes in atmospheric conditions and long-term shifts in climate patterns. These effects highlight the interconnected nature of wildfires and climate change, underscoring the need for integrated approaches to manage and mitigate their impacts.

[25] Climate forcing is a factor that changes the Earth's temperature. It can make the planet warmer (like greenhouse gases) or cooler (like certain aerosols).

Section II

Wildfire management strategies

3 Strategies for wildfire management

Strategizing involves crafting a purposeful plan to navigate complex and uncertain landscapes. In the realm of wildfire management, this art is crucial to channel resources — personnel, equipment, and funding — to where they will achieve the greatest effect. By carefully directing these resources, strategizing enhances the efficiency of wildfire management efforts, ensuring that every action has a significant impact.

Central to this process is the evaluation of risks, examining elements such as vegetation, weather, and historical fire patterns. This thoughtful analysis enables targeted interventions, such as controlled burns, firebreaks, and the creation of defensible spaces, which serve to mitigate the impact of wildfires. When a wildfire erupts, a well-devised strategy facilitates a cohesive response, encompassing evacuation plans, communication, and containment efforts, all executed with precision and timeliness.

Looking to the future, strategizing also entails preparing for resilience through land use planning, community education, and policy development. This forward-thinking approach helps manage costs by focusing on the most effective measures and minimizing the need for costly emergency interventions. Involving local communities in fire prevention further amplifies the effectiveness of operations, and is often a key component of wildfire management strategies.

3.1 WHAT IS MANAGEMENT?

Management is the art of orchestrating resources, activities, and strategies to achieve specific goals with maximum efficiency[1] and effectiveness[2]. In the realm of wildfire management, this art takes on a critical role, to ensure that society and nature are saved from the perils of wildfires, and also to make certain that we are able to circumvent threats in the most efficient and effective manner, parsimonious with regards to time, effort, and money.

[1]Efficiency refers to the ability to accomplish a task with the least amount of wasted resources, such as time, effort, or materials. In various contexts, it means getting the best possible outcome while minimizing inputs or costs. For example, in business, it might involve maximizing productivity, while in engineering, it could relate to optimizing processes or energy use.

[2]Effectiveness is the degree to which something successfully produces the desired result or outcome. It focuses on achieving goals, regardless of the resources used.

This necessitates a systematic and well-structured approach. Management provides the essential framework necessary to coordinate firefighting teams, equipment, and financial resources, ensuring that each effort contributes significantly to the containment and extinguishment of the fire.

This involves a thorough understanding of risks — and thus evaluating fire-prone areas, scrutinizing weather patterns, and monitoring vegetation are essential tasks of wildfire management. Such foresight allows for the implementation of preventative measures like firebreaks and controlled burns, addressing threats even before they spiral out of control.

Wildfire management also necessitates a flexible, adaptable, dynamic, and responsive management strategy. As the flames shift and the conditions evolve, management must respond swiftly, reallocating resources and adjusting tactics in real-time to ensure that efforts remain effective and impactful. Visionary management also encompasses long-term planning, integrating land use policies, community education, and comprehensive fire management strategies to build resilience against future fires. Judicious financial management ensures that resources are deployed wisely, minimizing economic strain on society and enhancing long-term sustainability.

3.2 THE DEMING CYCLE IN WILDFIRE MANAGEMENT

In wildfire management, the Deming cycle, or Plan-Do-Check-Act (PDCA) cycle, is an essential framework for effective response and its continuous improvement (Fugate et al., 2014). The Deming cycle is an iterative model that guides wildfire management teams through a process of ongoing refinement, ensuring that every phase of their efforts is carefully evaluated and enhanced.

The cycle begins with the **Plan** phase, where managers establish objectives and develop strategies to address the unpredictable nature of wildfires. This entails forecasting potential threats, identifying high-risk areas, and devising preventative measures such as controlled burns and strategic firebreaks. Comprehensive plans are meticulously crafted for resource allocation and response, with the objective of mitigating risks as expeditiously and cost-effectively as feasible.

The **Do** phase follows, where these plans are put into action. This stage encompasses the execution of the predetermined fire management techniques, encompassing those specifically designed for wildfire prevention, detection, suppression, and post-fire management.

In the **Check** phase, the focus shifts to reflection and analysis. Data are collected and reviewed to assess the effectiveness of various operations conducted during the Do phase. This analysis evaluates whether the strategies were able to achieve their intended outcomes and whether these outcomes were attained within the specified time and cost constraints. Insights gained from this analysis provide valuable information on both successes and areas for improvement.

Finally, the **Act** phase is where adjustments are made in the plan for the next cycle, based on the insights from the Check phase. Successful practices are standardized, while those that did not meet expectations are revised. This ongoing loop of evaluation and adaptation ensures that future wildfire management efforts become more efficient and effective, enhancing the ability to protect communities and landscapes while minimizing costs.

By integrating the Deming cycle, wildfire management teams adopt a dynamic process of continuous learning and improvement. This approach not only enhances their response capabilities but also fosters resilience over time, resulting in progressively more thoughtful and effective management of wildfires. We typically employ parallel Deming cycles for wildfire prevention, detection, suppression, and post-fire management operations. Let us now delve into these operations in greater detail.

3.3 PREVENTION

Wildfires pose a significant threat to both natural ecosystems and human communities, necessitating a comprehensive approach to prevent them. This entails the implementation of a comprehensive range of strategies to mitigate the risk and impact of wildfires, even before they occur. We will now delve into some of these strategies.

3.3.1 EDUCATION AND AWARENESS

Effective wildfire prevention begins with education and raising awareness of all stakeholders (Hesseln, 2018). Public outreach campaigns are instrumental in informing communities about the dangers of wildfires and the steps they can take to prevent them (Brenkert-Smith et al., 2013; Crow et al., 2015). These campaigns often include informational brochures, community workshops, and media advertisements, all designed to instill a sense of urgency and responsibility regarding fire safety.

Training programs are conducted for staff and community members to apprise them about the state-of-the-art wildfire detection, suppression, and post-wildfire management methods. Mock exercises are often a key component of these training programs (Stephens et al., 2023).

Schools play a pivotal role in disseminating wildfire education (Marquette, 2023). Incorporation of fire safety in the curriculum ensures that young people understand the risks and preventative measures associated with wildfires (Restaino et al., 2024). By teaching children about fire safety, we equip them with the knowledge to promote good practices at home, fostering a culture of awareness from an early age, something that lingers on as they become committed adults.

The digital age has massively aided education and awareness by providing new avenues in the form of online resources (Chauhan and Hughes, 2017). Websites and apps offer near-real-time alerts, prevention tips, and educational materials, making it easier for individuals to stay informed and take requisite measures.

3.3.2 FUEL MANAGEMENT

One of the primary strategies for wildfire prevention is the management of the fuel that wildfires consume. This involves reducing the quantity of combustible materials present in forests and surrounding properties.

Controlled burns, also known as prescribed fires, are a key method in this process (Wagle and Eakle, 1979). These fires are deliberately ignited under controlled conditions to consume excess vegetation and reduce the overall fuel load. While the idea of starting fires to control fires might seem counterintuitive, these controlled burns help prevent larger, uncontrollable wildfires.

Mechanical thinning is another technique used to manage fuel (Huggett Jr et al., 2008). By using machinery to remove excess trees and brush, we can decrease the density of vegetation, thereby reducing the fuel available for a wildfire to spread.

In some areas, controlled grazing by livestock is also employed as a method to manage vegetation (Donovan et al., 2022; Taylor Jr, 2006). Grazing helps to keep vegetation in check and reduce fuel loads, though it requires careful management to avoid overgrazing.

Furthermore, regular debris removal from forest floors and surrounding residential areas is paramount. Dead trees, leaves, and other flammable materials can readily ignite and facilitate the spread of wildfires, making their removal an indispensable component of fuel management strategies.

3.3.3 BUILDING CODES AND DESIGN

Robust building codes and their implementation can greatly help in the creation of defensible spaces (Syphard et al., 2014). Design and construction of buildings significantly influence their ability to withstand wildfires (Intini et al., 2020; McWethy et al., 2019). Adhering to fire-resistant building codes is a critical preventive measure. Structures built with materials such as metal roofing, stucco, and fire-resistant wood are less likely to catch fire compared to those made with more flammable materials.

Ember-resistant vents are another important design feature (Hakes et al., 2017). These vents prevent embers from entering attics or other vulnerable spaces, where they can ignite fires. Creating a defensible space around properties is also essential. This involves maintaining a cleared area of vegetation and using fire-resistant landscaping to create a buffer zone that can slow or stop the spread of a wildfire (Ondei et al., 2024).

3.3.4 FIREBREAKS AND BARRIERS

Creating physical barriers is a practical approach to controlling wildfire spread (Zong et al., 2021). Firebreaks are gaps in vegetation or other combustible materials designed to act as barriers to slow or halt the advance of wildfires (Awadhiya,

2017; Cui et al., 2019a). These can be natural features like rivers or rocky areas, or man-made clearings.

Buffer zones around high-risk areas, such as the interface between urban and wildland areas, are also crucial. These zones help to control the spread of fires by providing a space with reduced vegetation and fuel, and are often referred to as fuelbreaks (Ascoli et al., 2020; Wang et al., 2021).

In some cases, physical barriers such as walls or ditches are constructed to prevent fires from reaching critical areas (Awadhiya, 2017). These barriers can help protect homes, infrastructure, and other valuable assets.

3.3.5 COMMUNITY PLANNING

Effective wildfire prevention requires thoughtful community planning (Haines et al., 2008). Zoning regulations can play a significant role in minimizing wildfire risk by restricting development in high-risk areas (Mockrin et al., 2020). By controlling where and how development occurs, communities can reduce their vulnerability to wildfires (McWethy et al., 2019).

Evacuation plans are another critical component of community planning (Toledo et al., 2018). Developing and regularly practicing these plans ensures that residents can evacuate quickly and safely when a wildfire threatens. Infrastructure design is also important (Gonzalez-Mathiesen and March, 2018; Intini et al., 2020; Zehra and Wong, 2024); roads, bridges, and access routes must be designed to accommodate firefighting equipment and ensure that evacuation routes are clear and accessible.

3.3.6 FIRE DETECTION AND MONITORING

Early detection and monitoring are essential for effective wildfire management. Satellite technology has revolutionized fire detection (Davies et al., 2008), providing real-time data on fire activity, weather conditions, and vegetation health. This information allows for early intervention and better management of wildfire risks.

Traditional fire watch towers, staffed with observers or equipped with cameras, continue to play a role in detecting smoke and fires early (Zhang et al., 2020). Additionally, drones have also emerged as valuable tools for monitoring large areas, assessing fire behavior, and even delivering supplies when necessary (Afghah et al., 2019; Aydin et al., 2019).

3.3.7 BURN BANS AND OTHER REGULATIONS

Regulating activities that can spark wildfires is a crucial preventive measure. Implementing and enforcing burn bans during high-risk periods (especially those with dry and windy conditions) can significantly reduce the number of accidental ignitions (Hoang et al., 2013). Fireworks, which can easily cause wildfires (Vachula et al., 2023), are often subject to strict regulations or outright bans in high-risk areas.

Penalties imposed for violations of fire safety regulations serve as a deterrent and facilitate compliance. By holding individuals accountable for illegal burning and other risky behaviors, communities can reduce the likelihood of accidental wildfires (Chas-Amil et al., 2015).

3.3.8 RESEARCH AND INNOVATION

Ongoing research and technological innovation are vital for advancing wildfire prevention strategies (Xu et al., 2024). Studies on fire behavior help improve prediction models and inform better firefighting strategies (Albini, 1976b). Research into the impacts of climate change on wildfire frequency and intensity allows for the development of adaptive management strategies that address evolving risks (Dupuy et al., 2020; Williams et al., 2019).

New technologies, such as advanced fire retardants and improved firefighting gear, continue to enhance wildfire prevention and response capabilities (Carballo-Leyenda et al., 2017; Lu et al., 2023). Innovations like remote sensing technologies and improved data analysis tools further support efforts to manage and mitigate wildfire risks.

In summary, wildfire prevention is a multifaceted endeavor that necessitates a combination of educational initiatives, strategic planning, and technological advancements. By integrating these approaches and continuously adapting to evolving challenges, we can effectively manage and mitigate the risks associated with wildfires, safeguarding individuals and the environment.

3.4 DETECTION

Effective wildfire management hinges on early detection, a crucial factor that can significantly influence the outcome of firefighting efforts and reduce damage to ecosystems and communities. In an isotropic area[3], a wildfire will expand and extend in all directions at a continuous rate, as depicted in figure 3.1. This expansion of the wildfire will increase the size of the "fire front," and the difficulty of tackling and controlling the wildfire will increase with the time elapsed since its inception. While actual wildfires are influenced by factors such as the availability and quality of fuel, wind speed and direction, and the topography of the area, this general principle will continue to apply: the later we arrive at the site of the wildfire, the more challenging it will be to extinguish it. This section delves into various methods and technologies employed to detect wildfires swiftly and accurately, emphasizing the significance of advancements in this field in enhancing fire management and response.

[3]An isotropic area refers to a region in which properties are uniform in all directions.

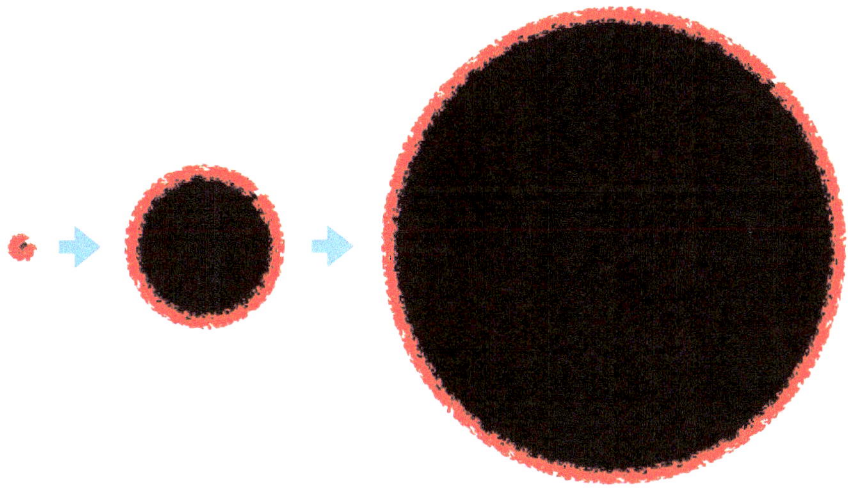

FIGURE 3.1: A schematic of the expansion of wildfire front with time.

3.4.1 TRADITIONAL DETECTION METHODS

Historically, wildfire detection relied on a combination of human observation and physical infrastructure. Fire watch towers in the forms of spar trees, tree platforms, crow nests, platform towers, and lookout cabins, strategically placed in vulnerable areas (Awadhiya, 2018), and manned by people with plane tables, alidades, binoculars, telescopes, and sometimes crank telephones, have long been a cornerstone of fire detection (Berryoung, 2021).

These towers are staffed with trained observers who monitor extensive landscapes for indications of wildfires in the form of light or smoke plumes, enabling early identification of fires, which is crucial for initiating a prompt response. Despite their effectiveness, fire watch towers have limitations, such as coverage gaps and susceptibility to weather conditions that can impair visibility. Consequently, the utilization of modern technologies for wildfire detection is essential.

3.4.2 TECHNOLOGICAL ADVANCES IN DETECTION

With the advent of technology, wildfire detection has become increasingly sophisticated. Satellite technology has revolutionized the field by providing real-time monitoring of large areas from space. Satellites equipped with thermal imaging sensors possess the ability to discern the heat signatures of wildfires, even amidst the presence of clouds and smoke. This capability enables the identification of fire hotspots that may have eluded detection through conventional methods.

Remote sensing satellites, such as those operated by NASA and other space agencies, offer comprehensive coverage and can track fire progression, assess fire intensity, and monitor changes in vegetation that may indicate increased fire risk (Chuvieco et al., 2020; Wooster et al., 2021). This data is invaluable for both early detection and ongoing fire management.

Another technological innovation is the use of drones for wildfire detection and monitoring (Allison et al., 2016; Mohapatra and Trinh, 2022). Drones equipped with high-resolution cameras and thermal sensors can effectively traverse challenging terrain and provide real-time imagery of fire conditions. They are particularly valuable for surveying fire perimeters, assessing damage, and gathering data in remote or hazardous areas where human presence may be hazardous.

3.4.3 ADVANCED MONITORING SYSTEMS

In addition to satellites and drones, a variety of advanced monitoring systems are employed to detect wildfires. One such system is the network of fire detection cameras. These cameras are strategically placed in high-risk areas and use optical and infrared technology to detect smoke and heat (Heyns et al., 2019). The data collected by these cameras is transmitted to control centers in real-time, where it is analyzed for signs of wildfire activity.

Weather stations also play a critical role in wildfire detection. Equipped with sensors to measure temperature, humidity, wind speed, and other meteorological factors, these stations provide essential data that can help predict fire behavior and identify conditions conducive to wildfire ignition (Naganathan et al., 2016). Some modern weather stations are integrated with fire detection systems, offering a comprehensive view of fire risk and enabling early intervention.

3.4.4 MACHINE LEARNING AND ARTIFICIAL INTELLIGENCE

Recent advancements in machine learning and artificial intelligence (AI) have further enhanced wildfire detection capabilities (Joshi et al., 2024). Artificial intelligence algorithms can process extensive data from diverse sources, such as satellite imagery, camera feeds, and weather data, to discern patterns and forecast wildfire occurrences. Machine learning models can be trained to recognize the visual and thermal signatures of wildfires, improving the accuracy and speed of detection (Jain et al., 2020).

AI systems can also process data in real-time, providing timely alerts to firefighters and emergency responders. These systems can also prioritize alerts based on the severity and potential impact of detected fires, ensuring that resources are allocated efficiently (Bot and Borges, 2022).

3.4.5 COMMUNITY AND CROWD-SOURCED REPORTING

In addition to technological solutions, community and crowd-sourced reporting have become valuable tools in wildfire detection. Many regions have established platforms that allow residents to report smoke sightings or fire activity (Bogdos and

Manolakos, 2019; Zhong et al., 2016). These reports are often verified by fire agencies and used to supplement official detection systems.

Social media platforms and mobile apps enable the public to share information about wildfire sightings, which can be rapidly analyzed and used to improve situational awareness (Slavkovikj et al., 2014). While crowd-sourced data requires careful validation to ensure accuracy, it can provide critical information that complements other wildfire detection methods.

3.4.6 INTEGRATED DETECTION SYSTEMS

The most effective wildfire detection strategies often involve integrating multiple detection methods to create a comprehensive monitoring network (Ambrosia et al., 1998). Combining satellite imagery, drone surveillance, fire detection cameras, weather stations, and AI analyses provides a multi-layered approach to early detection.

For example, data from satellites can be cross-referenced with information from ground-based cameras, weather stations, and social media feeds to confirm the presence of a wildfire and its extent, while also assessing its potential impact and deciding on the best way to tackle it. This integrated approach makes detection systems more robust and capable of responding to a range of wildfire scenarios.

3.4.7 CHALLENGES AND FUTURE DIRECTIONS

Despite the advancements in wildfire detection, several challenges persist. Detecting fires in remote or inaccessible areas remains challenging, and the accuracy of detection systems can be compromised by environmental factors such as smoke, clouds, and terrain. Furthermore, the substantial amount of data generated by contemporary detection systems necessitates effective management and analysis to ensure timely responses.

Future developments in wildfire detection will likely prioritize enhancing the precision and reliability of existing technologies. Innovations in satellite and drone technology, coupled with advancements in artificial intelligence and machine learning, will continue to augment our detection capabilities. Collaborative efforts between technology developers, researchers, and firefighting agencies will be pivotal in addressing current challenges and advancing detection systems.

3.5 SUPPRESSION

Wildfire suppression is a complex and challenging endeavor involving a range of strategies and technologies designed to control and extinguish fires. Effective suppression necessitates not only the implementation of robust tactics and tools but also a coordinated effort involving various stakeholders, such as firefighting agencies, local communities, and governmental organizations. In this section, we delve into the methods and technologies employed in wildfire suppression, emphasizing the evolution of tactics, the role of contemporary technology, and the challenges encountered by firefighting teams.

3.5.1 FUNDAMENTAL SUPPRESSION STRATEGIES

Wildfire suppression strategies can be broadly categorized into direct and indirect methods. Direct suppression involves actively engaging with the fire to extinguish it or control its spread. It happens in close proximity to the wildfire and is especially useful for smaller fires. Direct suppression includes using hand tools, hoses, and fire retardants to attack the flames directly. Firefighters work to create firelines — cleared areas devoid of combustible materials — around the fire to contain and suppress its advance. Water is applied to cool the fire and extinguish flames, while fire retardants, typically a combination of water and chemicals, are employed to diminish the flammability of vegetation and mitigate the fire's spread. Firefighters may also utilize helicopters or fixed-wing aircraft to drop these retardants on inaccessible areas or to reinforce firelines.

Indirect suppression, on the other hand, involves strategies that aim to prevent the fire from spreading further. It is frequently necessary for moderate and high-intensity wildfires that pose a risk to personnel working in close proximity. Indirect suppression entails establishing firebreaks at a distance and employing controlled burns to eliminate fuel in the path of the wildfire.

3.5.2 TECHNOLOGICAL INNOVATIONS IN SUPPRESSION

The role of technology in wildfire suppression has grown significantly, enhancing the effectiveness and efficiency of firefighting efforts. A key technological advancement is the development of aerial firefighting tools (Xanthopoulos et al., 2020). Aircraft equipped with water tanks or fire retardant dispensers play a crucial role in battling large wildfires. These aircraft can drop significant amounts of water or retardant quickly over large areas, providing critical support to ground crews.

Drones have also become an invaluable asset in wildfire suppression. Utilizing thermal imaging cameras and high-resolution sensors, drones can provide real-time data on fire behavior, assist in mapping fire perimeters, and identify hotspots that necessitate immediate attention. They offer a comprehensive aerial perspective of the fire, which is indispensable for devising effective suppression strategies and safeguarding the well-being of firefighting personnel.

Another significant advancement is the use of remote sensing technologies, which provide detailed information on fire intensity, smoke dispersion, and vegetation health. Satellite data and infrared sensors can monitor fire activity from space, offering insights that help guide suppression efforts and resource allocation.

3.5.3 FIREFIGHTING EQUIPMENT

Firefighting equipment has evolved to meet the demands of modern wildfire suppression. Fire engines, equipped with high-capacity water tanks and powerful hoses, are essential for direct suppression efforts. These engines are engineered to operate in challenging environments and are frequently accompanied by support vehicles that transport additional supplies and equipment.

Hand tools such as shovels, rakes, and axes, along with mechanized vegetation cutters and earth-moving machines, are fundamental to creating firelines and clearing vegetation. These tools facilitate firefighters' close proximity to the fire, enabling them to execute critical tasks for containment.

Personal protective equipment (PPE) for firefighters has also seen advancements, including heat-resistant clothing, helmets, and breathing apparatus. These protective measures are meticulously designed to safeguard firefighters from the intense heat, smoke, and hazardous conditions that are commonly encountered during suppression operations.

3.5.4 COORDINATION AND COMMUNICATION

Effective wildfire suppression hinges on seamless coordination and communication among diverse teams and agencies. Firefighters, emergency responders, and support personnel must collaborate harmoniously to execute suppression strategies and adapt to evolving fire conditions.

To this end, Incident Command Systems (ICS) are used to organize and manage firefighting efforts (Chang, 2017). These systems establish clear lines of authority and ensure that resources are allocated efficiently. The National Incident Management System (NIMS) (Agency, 2017) and the Incident Command System (ICS) (McAllister et al., 2020) are examples of frameworks that facilitate coordinated response efforts.

Communication technologies, including radios, satellite phones, and GPS systems, are critical for maintaining real-time contact between teams and coordinating efforts across vast and often challenging landscapes. These technologies facilitate effective command and control, ensuring that all personnel are adequately informed and can promptly respond to emerging developments.

3.5.5 CHALLENGES AND FUTURE DIRECTIONS

Wildfire suppression faces numerous challenges, including unpredictability due to weather, wind patterns, and vegetation. Rugged terrain impedes access and complicates suppression operations, necessitating the utilization of aerial firefighting equipment and drones. Resource constraints, including personnel, equipment, and funding, can strain response efforts during extensive wildfire events.

Technological advancements, including autonomous firefighting drones, advanced fire retardants, and improved predictive modeling, will enhance wildfire combat.

Research on new fire retardants aims to improve effectiveness and reduce environmental impact. Data analytics and machine learning will continue to enhance fire behavior predictions, facilitating more effective planning and resource allocation.

Collaboration among scientists, technology developers, and firefighting agencies is paramount to the development and implementation of novel solutions. Integrating emerging technologies into existing suppression strategies will further improve wildfire management and mitigation, thereby reducing wildfire devastation and safeguarding communities and natural landscapes.

3.6 POST-WILDFIRE MANAGEMENT

Post-wildfire management is a critical phase in the lifecycle of a wildfire, focusing on recovery, restoration, and rebuilding efforts following the containment and extinguishment of the fire. This section delves into the diverse methodologies employed in post-wildfire management, encompassing ecological restoration, community recovery, and infrastructure reconstruction. Through an examination of the strategies and challenges associated with these approaches, we gain insights into effective strategies for supporting ecosystems and communities in the aftermath of wildfires.

3.6.1 ECOLOGICAL RESTORATION

One of the primary goals of post-wildfire management is to restore the affected ecosystems to their pre-fire conditions. Wildfires have profound impacts on soil, vegetation, and wildlife, making restoration efforts essential for the health of the environment.

Soil rehabilitation: Wildfires can lead to soil erosion, loss of nutrients, and increased runoff. To address these issues, soil rehabilitation techniques are employed (Robichaud, 2009). This encompasses stabilizing soil through the implementation of erosion control measures, such as silt fences, straw mats, and re-seeding with native plants. These practices effectively prevent further erosion and foster the restoration of soil health.

Vegetation restoration: The restoration of vegetation involves replanting native species that were damaged or destroyed by the wildfire (Long et al., 2014). This process typically commences with identifying the species most suitable for the post-fire environment. Restoration can be achieved through artificial, assisted natural, or natural regeneration. In artificial regeneration, replanting or seeding operations are conducted, usually with the assistance of local communities. In assisted natural regeneration, seedlings that have naturally emerged are safeguarded from herbivory, diseases, pests, and calamities, thereby enhancing their chances of growing into trees. In natural regeneration, the protection of remaining vegetation is prioritized, along with the creation of conditions conducive to natural seed dispersal and growth. Consequently, the subsequent generation of plants emerges through natural processes.

Wildlife recovery: Wildfires can displace wildlife and destroy their habitats. Post-wildfire management encompasses initiatives aimed at supporting wildlife recovery. These efforts typically involve the provision of habitat structures, such as artificial nests or shelters, and monitoring wildlife populations to evaluate the progress of recovery. Restoration projects may also prioritize the creation of wildlife corridors to facilitate the reconnection of fragmented habitats and enhance animal movement. In extreme circumstances, translocation of individuals from priority species may be considered to establish a founding population.

3.6.2 COMMUNITY RECOVERY

The impact of wildfires on communities can be devastating, affecting homes, businesses, and infrastructure. Effective community recovery involves addressing immediate needs and planning for long-term rebuilding and resilience.

Emergency relief and support: Immediately following a wildfire, emergency relief efforts are crucial for providing temporary housing, medical care, and basic necessities to those affected. Organizations and agencies are actively engaged in providing essential support to individuals and families affected by the fire. This support encompasses the distribution of supplies, financial assistance, and counseling services aimed at assisting those coping with the trauma and disruptions caused by the incident.

Rebuilding and reconstruction: The process of rebuilding homes and infrastructure involves careful planning and coordination. This includes assessing the damage, securing permits, and ensuring that new constructions meet updated building codes and fire-resistant standards (also known as the build back better paradigm). Rebuilding efforts frequently prioritize enhancing resilience to future wildfires by incorporating fire-adaptive designs and materials.

Community engagement and support: Engaging with affected communities is essential for effective recovery. This entails involving residents in decision-making processes, ensuring transparent communication regarding recovery plans, and supporting community-led initiatives. Recovery programs may also incorporate workshops and educational events to assist residents in preparing for future wildfires and bolstering community resilience.

3.6.3 INFRASTRUCTURE RESTORATION

Restoration of infrastructure damaged by wildfires is a complex process that requires coordination among various stakeholders and agencies. This encompasses the restoration or replacement of damaged roads, utilities, and public facilities.

Roads and transportation: Wildfires can damage or destroy roads, bridges, and other transportation infrastructure. Restoring and reconstructing these critical assets entails evaluating structural damage, removing debris, and guaranteeing the safety and accessibility of transportation corridors. In certain instances, infrastructure enhancements may encompass improving road design to enhance resilience against future wildfires.

Utilities and services: Restoration of utilities such as water, electricity, and gas is a priority in post-wildfire recovery. This entails repairing damaged lines, replacing equipment, and ensuring the safe and efficient restoration of utility services. Utility companies must collaborate closely with emergency responders and local authorities to address service disruptions and prioritize restoration efforts.

Public facilities: Rebuilding public facilities such as schools, parks, and community centers is an important aspect of recovery. This process entails assessing the extent of damage, devising plans for repairs or reconstruction, and ensuring that facilities adhere to the most recent safety and resilience standards.

3.6.4 MONITORING AND EVALUATION

Effective post-wildfire management requires ongoing monitoring and evaluation to assess the progress of recovery efforts and identify areas for improvement. This entails monitoring the outcomes of ecological restoration initiatives, assessing the efficacy of community recovery programs, and evaluating the state of rehabilitated infrastructure.

Ecological monitoring: Monitoring the progress of ecological restoration involves assessing vegetation regrowth, soil health, and wildlife populations. This data serves as a valuable resource in assessing the efficacy of restoration initiatives and formulating adaptive management strategies.

Community Impact Assessment: Evaluating the impact of recovery programs on affected communities involves gathering feedback from residents, assessing changes in quality of life, and measuring the success of rebuilding and support initiatives. This information helps to identify successes and challenges in the recovery process and inform future strategies.

Infrastructure performance: Monitoring the performance of repaired or rebuilt infrastructure involves assessing its functionality, safety, and resilience. This encompasses conducting inspections, assessing repair quality, and verifying that infrastructure aligns with community requirements.

3.6.5 CHALLENGES AND FUTURE DIRECTIONS

Post-wildfire management presents several challenges, including limited resources, the complexity of recovery efforts, and the need to balance immediate and long-term needs. Ensuring that recovery efforts are equitable, effective, and sustainable necessitates meticulous planning, coordination, and collaboration among various stakeholders.

Future directions in post-wildfire management may involve integrating novel technologies and approaches to enhance recovery efforts. Innovations such as remote sensing for monitoring ecological recovery, advanced building materials for improved resilience, and data analytics for assessing community impact can contribute to more effective and efficient recovery processes.

Furthermore, fostering collaboration between governmental agencies, non-profit organizations, and community groups will be paramount in addressing the multifaceted challenges of post-wildfire recovery. By synergizing diverse expertise and resources, we can support resilient communities and maintain healthy ecosystems in the aftermath of wildfires.

4 Wildfire prevention

Wildfires pose a significant and escalating threat to both natural ecosystems and human settlements. Uncontrolled fires can rapidly devastate extensive landscapes, causing property damage, disrupting wildlife habitats, and compromising air quality. Consequently, the imperative is to promptly extinguish wildfires and, ideally, prevent their occurrence and subsequent spread.

Preventing wildfires demands a multifaceted approach, integrating responsible land management practices, community education, and the use of fire-resistant materials (Hesseln, 2018). Effective land management includes creating firebreaks, clearing combustible vegetation, and conducting controlled burns to reduce fuel loads. Community education plays a crucial role in fostering awareness of fire safety and the importance of maintaining defensible spaces around properties. By engaging the public in these preventive measures, we seek to bolster collective resilience against wildfire risks. Furthermore, adopting fire-resistant construction materials and adhering to safe outdoor practices contribute to mitigating ignition hazards.

In this chapter, we will explore various strategies employed to prevent wildfires.

4.1 NEED FOR DATA AND DATA ANALYSIS

Data is fundamental to wildfire prevention, embodying the principle that "what cannot be measured cannot be managed." By harnessing the power of data, we can refine our strategies and make informed decisions to better protect both natural landscapes and human communities from the destructive impacts of wildfires. This also means that we need more and more data and data analyses for effective wildfire management (Bowman, 2018).

We need data for predictive modeling and risk assessment. Historical fire records, together with current weather patterns and vegetation conditions, can help pinpoint areas at high risk and to anticipate potential fire behavior (Awadhiya, 2018), thus enabling the movement of men and materials to those locations where they are most needed. Advanced simulations derived from data can effectively visualize various fire scenarios and implement the most effective preventive measures, including controlled burns and strategic vegetation management in high-risk areas.

Besides historical data, we also need real-time data for monitoring and early detection to prevent the spread of wildfires once they have started. Satellite imagery, aerial surveys, and ground-based sensors provide a continuous stream of information about weather conditions, vegetation availability, and fire activity, allowing for early identification of emerging fires or hotspots, and facilitating swift action to contain them before they escalate into uncontrollable blazes.

Prior planning and resource allocation are similarly dependent on data-driven tools. Geographic Information Systems (GIS) and other analytical platforms are used to create detailed maps of fire-prone areas, infrastructure, and available resources, supporting decision-making processes, allocation of resources, and the development of evacuation plans.

Public awareness and education also benefit from a data-informed approach. Data — such as the percentage of houses with and without defensible spaces that burn in wildfires — underline and highlight the salient concepts of educational campaigns through availability heuristics (Keller and Block, 1997), and thus help to foster a culture of preparedness (Jakes et al., 2003; Sturtevant and McCaffrey, 2006). This initiative encourages responsible behavior, including the proper disposal of flammable materials and the maintenance of defensible space around properties. By implementing these measures, we can effectively reduce the overall wildfire risk.

Data remains of paramount importance even after the wildfire. It serves as the foundation for assessing the fire's impact, evaluating the effectiveness of response efforts, and drawing valuable lessons from the event. Post-fire analyses inform future strategies, enhance response protocols, and bolster community resilience through inputs to the Deming's cycle. All these pave the way for increasingly more effective wildfire management in the future.

For these reasons, we have been developing and advancing more ways to capture and analyze data for wildfire management purposes (Artés et al., 2019).

4.2 FROM DATA TO INFORMATION

Data (singular datum) includes raw, unprocessed facts and figures collected from various sources. Data can be quantitative or qualitative and may take various forms, such as numbers, text, or measurements. When examined directly, data often presents itself as a disorganized collection of numbers and letters, lacking any discernible pattern or meaning. In essence, data alone does not convey inherent significance or context; it is simply a compilation of distinct elements.

To make sense of data, we have moved into the domain of **information**. Information is what emerges when data is processed, organized, and interpreted to provide context and meaning. It involves analyzing data to uncover patterns, relationships, and insights, which can then be used to make informed decisions. While data represents the building blocks of decision-making, information is the structured and meaningful output that supports understanding and decision-making.

The journey from data to information involves several critical steps, each adding some value to the raw data (Van Der Aalst, 2016). The process begins with **data collection**, where raw data is gathered from various sources such as sensors, surveys, or databases. This step is fundamental, as the quality and relevance of the collected data directly impact the effectiveness of subsequent analyses.

After collection, data is put through the operations of **data cleaning and preprocessing**. Raw data often contains inaccuracies, inconsistencies, or irrelevant details that need to be addressed. Data cleaning involves correcting errors, removing duplicates, and managing missing values. Preprocessing might include standardizing formats and normalizing values to ensure the data is consistent and suitable for analyses.

Following preprocessing, the data is **integrated** if it originates from multiple sources. Integration involves combining and aligning data from different origins, ensuring a cohesive dataset that provides a comprehensive view. This step is crucial for creating a unified data set that accurately reflects the full scope of the information.

With clean and integrated data, the next step is **data analysis**. This phase employs statistical techniques, algorithms, or models to discern patterns, trends, and relationships within the data. The analysis can be descriptive, exploratory, inferential, predictive, or prescriptive, depending on the objectives. Descriptive analysis provides a summary of the data's fundamental characteristics, while exploratory data analysis employs visualizations to unveil inherent patterns. Inferential analysis draws conclusions about populations based on samples, and predictive analysis forecasts future trends. Prescriptive analysis recommends specific actions based on the insights gained from the data.

After analysis, **data interpretation** is required to make sense of the results. This stage entails contextualizing the findings to comprehend their relevance and implications. Interpretation necessitates domain expertise to draw insightful conclusions and establish connections between the data and practical scenarios.

The next step is **information visualization**, where the analyzed data and findings are presented in visual formats such as charts, graphs, or maps. Visualization aids in the clear and effective communication of intricate information, ensuring its accessibility and comprehensibility to stakeholders.

Once visualized, the information is communicated through **reporting and communication**. This entails the compilation of reports, summaries, or presentations that elucidate salient insights and provide actionable recommendations. Effective communication ensures that the information is readily comprehensible and can serve as a guiding principle for decision-making processes.

The ultimate goal of transforming data into information is to support **decision-making**. In this stage, the insights derived from the data serve as the foundation for strategic planning, operational decisions, and policy development. By applying this information to real-world contexts, decision-makers can make informed choices that are grounded in evidence.

Finally, **feedback and iteration** play a crucial role in the process. This process entails monitoring and evaluating the outcomes of decisions made based on the information provided. This feedback facilitates the refinement of data collection methods, analysis techniques, and interpretation processes, resulting in continuous improvement in the quality and utility of the information.

In the context of wildfire prevention, transforming raw data into actionable information entails a meticulous process that commences with the collection of diverse data, including weather conditions, vegetation types, and historical fire records. This data undergoes rigorous cleaning and preprocessing to rectify inaccuracies and standardize formats, ensuring its reliability. Subsequently, the data is integrated to construct a comprehensive dataset that encompasses a holistic view of fire risk factors and other pertinent information.

Through meticulous analyses, distinct patterns and trends emerge, highlighting regions with elevated fire risk due to dry vegetation and formidable winds. This data is subsequently interpreted to comprehend its implications for matters such as wildfire risk. The insights are presented graphically through risk maps and dashboards. Effective communication of these findings to stakeholders facilitates informed decision-making, thereby leading to strategic preventive measures, including targeted vegetation management and enhanced emergency response protocols. Continuous feedback and iterative refinement further enhance the efficacy of wildfire prevention strategies over time. Each step contributes value, transforming raw data into insightful knowledge that supports comprehension and sound decision-making.

4.3 WILDFIRE HAZARD INDICES

We often require a single number to evaluate, understand, and predict the potential risk and severity of wildfires [Figure 4.1], drawing into it the essence of multitudes of environmental and climatic factors. This may be needed for simplicity and easy dissemination of actionable information to several stakeholders, some of whom may not have the time or resources to delve deep into what each and every factor represents. For such cases, we make use of wildfire hazard indices. These indices serve as indispensable tools in the field of fire management and environmental protection, providing a comprehensive assessment of wildfire risk. This information is paramount for effectively preparing and responding to wildfire threats. By comprehending and utilizing these indices, fire management agencies, policymakers, and communities can promptly anticipate fire behavior, implement mitigation measures, and mitigate the detrimental effects of wildfires.

4.3.1 COMPONENTS OF WILDFIRE HAZARD INDICES

Wildfire hazard indices integrate several key parameters to assess fire risk.

Weather conditions are fundamental to wildfire hazard assessment. Factors such as temperature, relative humidity, wind speed, and precipitation levels are closely monitored. Elevated fire risk can be attributed to high temperatures and low humidity, which dry out vegetation and make it more combustible. Conversely, high

FIGURE 4.1: Fire danger rating, a single value that depicts the likelihood of initiation of fire in a given area, based on current and recent weather patterns, fuel types and availability, and moisture.

humidity can reduce fire risk by increasing moisture content in the fuel. Wind speed is another critical factor; strong winds can rapidly spread fires by carrying embers to new locations, thereby increasing the fire's potential impact. Additionally, hot and dry winds can prime vegetation for burning, increasing the likelihood of a wildfire spreading swiftly. Katabatic winds blowing down from hills are especially notorious in this regard, and have played a pivotal role in the swift spread of Camp wildfire [Figure 1.1].

Secondly, **fuel characteristics** of the area are assessed. This encompasses the type, quantity, and condition of vegetation and other combustible materials present. Different types of fuel, such as grasses, shrubs, and trees, exhibit varying flammability and combustion characteristics. For instance, dry grasslands are more prone to ignition compared to moist, dense forests. The condition of the fuel, particularly its moisture content, significantly influences fire behavior.

Thirdly, **topography** influences how fires spread. The slope of the land and its aspect (the direction it faces) play pivotal roles in comprehending fire dynamics. Fires tend to spread more rapidly uphill due to the preheating of the fuel above them by the flames, facilitating their continued ascent. Additionally, the aspect of a slope influences fuel moisture levels. In the Northern Hemisphere, South-facing slopes receive more sunlight and, consequently, have drier fuels compared to North-facing slopes. Conversely, in the Southern Hemisphere, North-facing slopes receive more sunlight and, therefore, have drier fuels compared to South-facing slopes.

Additionally, **drought conditions** are a critical factor. Prolonged droughts can significantly reduce soil moisture and cause vegetation to dry out, enhancing its susceptibility to fire. Drought conditions also contribute to the accumulation of dead and dry fuel, further intensifying fire risk. This is particularly pronounced in regions with deciduous trees that shed their leaves during dry seasons as a moisture-saving mechanism.

Human factors also play a significant role in wildfire hazard assessment and are incorporated in wildfire hazard indices, especially since the majority of wildfires today — around 70 to 90% (Robinne and Secretariat, 2021) — are caused by human activities. Land use patterns, historical fire data, and human activities such as construction, land clearing, and recreational use significantly influence wildfire risk. Regions with elevated human activity or inadequate land management practices may be more prone to ignitions and encounter heightened difficulties in suppressing wildfires.

4.3.2 TYPES OF WILDFIRE HAZARD INDICES

Wildfire hazard indices come in various forms, each serving distinct purposes and applications.

Fire Danger Rating Systems [Figure 4.1] are among the most commonly used indices. These systems integrate data pertaining to weather conditions, fuel availability, and topographical features to generate an index that quantifies the likelihood of wildfire ignition and subsequent spread. The National Fire Danger Rating System (NFDRS) (Deeming, 1972) in the United States and the Canadian Forest Fire Danger Rating System (CFFDRS) (Stocks et al., 1989; Van Wagner et al., 1974) are well-known examples. These systems categorize fire danger into distinct levels, including low, moderate, high, very high, and extreme, thereby enabling critical and speedy decision-making in fire management.

The **Fire Weather Index (FWI)** is a component of the CFFDRS and focuses specifically on fire weather conditions (Van Wagner et al., 1974). This integrated system incorporates temperature, humidity, wind speed, and precipitation data to calculate fire danger levels. The Fire Weather Index generates a numerical value that quantifies the probability of fire ignition and its potential intensity, enabling fire managers to effectively assess current fire risk and make appropriate preparations.

Fire Behavior Prediction Models such as the BEHAVE system offer detailed forecasts of fire behavior based on real-time and predicted weather conditions, fuel types, and topography (Burgan, 1984). These models simulate diverse fire scenarios, offering valuable insights into potential fire spread patterns, intensity, and impacts. Consequently, they are indispensable for formulating fire suppression strategies and allocating resources effectively. They may also be employed to calculate wildfire hazard indices.

Remote Sensing Indices utilize satellite and aerial imagery to provide real-time data on vegetation moisture content, fuel conditions, and other relevant factors (Leblon, 2001). The Normalized Difference Vegetation Index (NDVI) (Rouse et al., 1974) is an example of such an index. NDVI measures the density of green vegetation and can assist in assessing the health and moisture content of vegetation over extensive areas. This information is paramount for monitoring changes in fuel conditions and identifying early indications of drought or fire risk. Such indices may be integrated into certain wildfire hazard indices.

4.3.3 APPLICATIONS OF WILDFIRE HAZARD INDICES

Wildfire hazard indices serve multiple purposes and are utilized by various stakeholders for distinct applications. Wildfire managers employ wildfire hazard indices to make well-informed decisions regarding resource allocation, fire suppression strategies, and public safety measures. By comprehending the current fire danger levels, fire managers can prioritize areas for intervention, allocate resources more efficiently, and implement strategies to contain and control fires effectively.

Communities utilize these indices to assess their susceptibility to wildfires and devise preventive measures. Strategies informed by hazard indices include creating defensible spaces around properties, conducting controlled burns to mitigate fuel loads, and developing comprehensive evacuation plans. This proactive approach can substantially diminish the risk of property damage and bolster community resilience.

Researchers employ hazard indices to analyze trends and patterns in wildfire occurrences, serving as a proxy that incorporates fire behavior, climate interactions, and land management practices. This research also contributes to the refinement of hazard indices, enhancing their accuracy and reliability over time.

Fire management policies and practices are developed by planners based on insights gained from wildfire hazard indices.

4.4 IDENTIFYING WILDFIRE-PRONE AREAS

Identifying wildfire-prone areas is a multifaceted process that necessitates a comprehensive understanding of various environmental, climatic, and human factors that contribute to wildfire risk. This process entails several systematic steps that are essential for the accurate assessment and prediction of regions susceptible to wildfires.

We often begin by collecting data about the place [Figure 4.2a] and time [Figure 4.2b] of historical wildfires (Awadhiya, 2022b). Given that the causal factors for wildfires are not anticipated to undergo significant shifts over the next few years, we can reasonably expect that regions identified as having the highest frequency of wildfires will persistently experience an elevated incidence of wildfires in the future. Similarly, seasons (or months) characterized by exceptionally high wildfire numbers will likely continue to generate a substantial number of wildfire alerts in the immediate future.

Just using historical information, we can construct fire maps [Figure 4.2c] that represent the propensity of an area to have wildfires based on historical wildfire data (Awadhiya, 2018). In the figure, forest compartments with no fire alerts in the past five years are depicted in green, while those with an increasing number of fire alerts are indicated by redder colors. A visual inspection of the figure reveals that the fringe areas of Satpura Tiger Reserve are predominantly red, while the central areas are predominantly green. This observation suggests that for effective wildfire management, a greater allocation of resources (including personnel, equipment, and watchtowers) should be directed towards the fringe areas compared to the central areas.

Certainly, we may encounter situations where an extensive wildfire in a specific year significantly reduces the fuel availability within the forest, thereby decreasing the likelihood of future wildfires. But by analyzing multi-year data, we can mitigate the influence of a single year's wildfire on a fire map. Furthermore, our understanding of the causal factors of wildfires serves as an additional check. If these factors have remained consistent over time, we can anticipate that the wildfire pattern will also be similar to previous years.

It is estimated that globally around 70 to 90% of wildfires are started by humans (Robinne and Secretariat, 2021). Thus, the presence of humans and human activities are good correlates of wildfire occurrence. When we plot the accessibility map of Satpura Tiger Reserve [Figure 4.2d] representing time taken to reach a spot from the nearest human habitation, we find that inaccessible areas and areas far from human habitations have fewer wildfires, and accessible areas close to human habitations have more wildfires. If there has not been a substantial shift in the distribution of human populations (for instance, through the expansion or relocation of habitations to newer areas), and no significant alteration in accessibility patterns (such as the construction of new roads or substantial changes in fuel prices and vehicle availability), we can anticipate that the fire map will provide a reasonably accurate representation of the current situation.

We can also explore changes in human activities, particularly changes in agriculture. In figure 4.2e, we observe near-real-time fire alerts in the SimplyFire application. The depicted area is the Sheopur district of India, with the district boundary outlined in blue. Forest areas are situated in the right portion of the district and are overlain with forest beat boundaries in green. Agricultural areas are located in the left portion of the district and do not contain green lines representing forest beats. Fire alerts from various satellites are depicted as red dots, pink dots, and fire symbols. It is evident that while agricultural and non-forest lands are saturated with fire signatures, there are minimal or no fire signatures in the forested areas. This

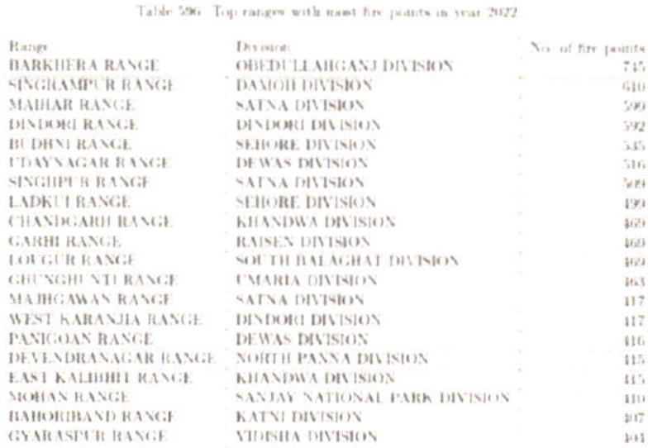

Table 5.96: Top ranges with most fire points in year 2022

Range	Division	No. of fire points
BARKHERA RANGE	OBEDULLAHGANJ DIVISION	745
SINGRAMPUR RANGE	DAMOH DIVISION	610
MAIHAR RANGE	SATNA DIVISION	599
DINDORI RANGE	DINDORI DIVISION	592
BUDHNI RANGE	SEHORE DIVISION	535
UDAYNAGAR RANGE	DEWAS DIVISION	516
SINGHPUR RANGE	SATNA DIVISION	509
LADKUI RANGE	SEHORE DIVISION	499
CHANDGARH RANGE	KHANDWA DIVISION	469
GARHI RANGE	RAISEN DIVISION	469
LOUGUR RANGE	SOUTH BALAGHAT DIVISION	469
GHUNGHUTI RANGE	UMARIA DIVISION	463
MAJHGAWAN RANGE	SATNA DIVISION	417
WEST KARANJIA RANGE	DINDORI DIVISION	417
PANIGOAN RANGE	DEWAS DIVISION	416
DEVENDRANAGAR RANGE	NORTH PANNA DIVISION	415
EAST KALIBHIT RANGE	KHANDWA DIVISION	415
MOHAN RANGE	SANJAY NATIONAL PARK DIVISION	410
BAHORIBAND RANGE	KATNI DIVISION	407
GYARASPUR RANGE	VIDISHA DIVISION	404

(a) Forest ranges with the highest number of wildfires in the year 2022.

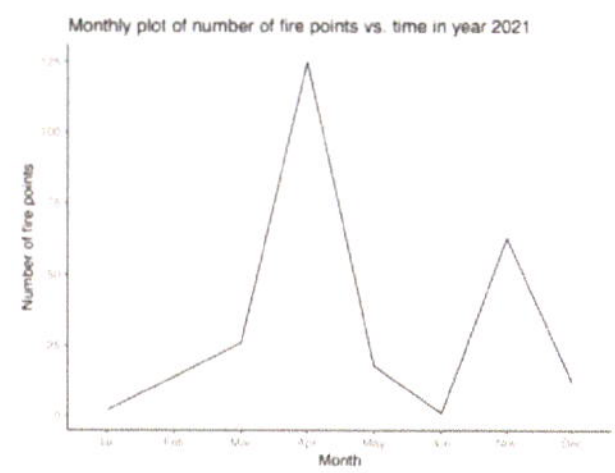

(b) Monthly plot of wildfire alerts in Gwalior for the year 2021.

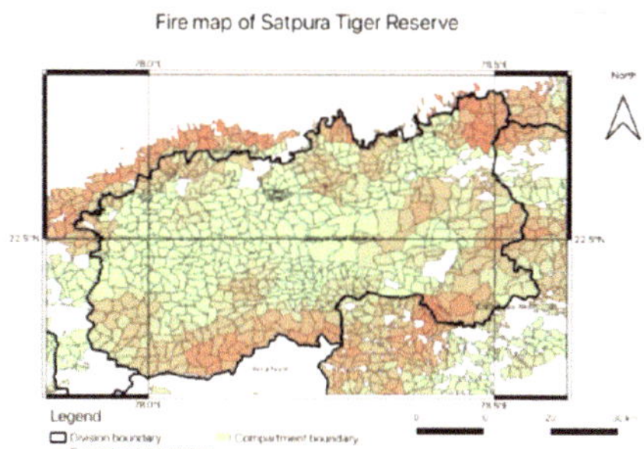

(c) Fire map of Satpura Tiger Reserve depicting wildfire alerts in the past 5 years.

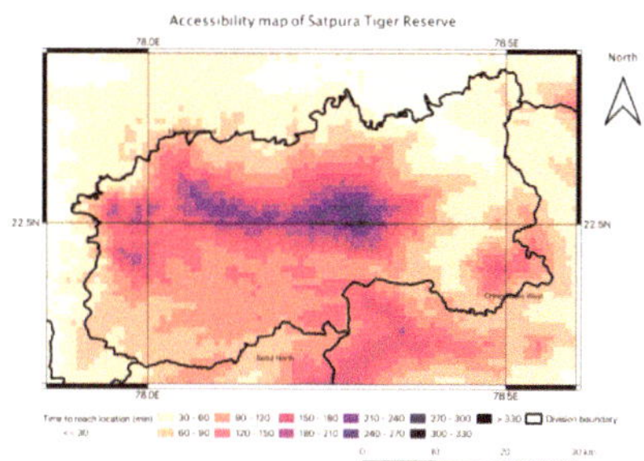

(d) Accessibility map of Satpura Tiger Reserve, with accessibility decreasing from yellow to blue.

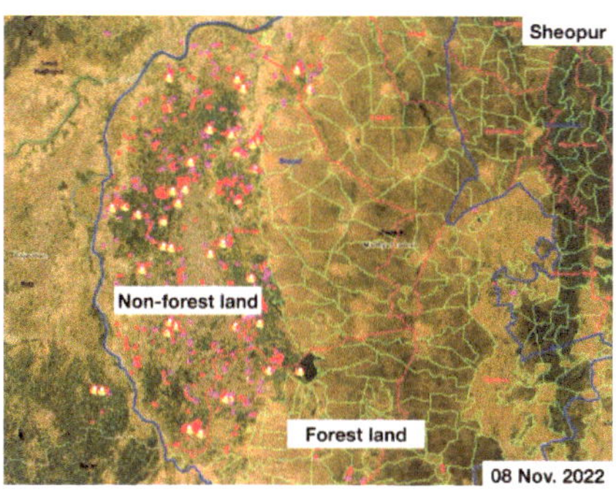

(e) Near-real time fire alerts in Sheopur district on 08 November 2022 visualized on SimplyFire.

(f) Near-real time fire alerts around Serra do Pardo National Park on 24 September 2024 visualized on SimplyFire.

FIGURE 4.2: Identification of fire-prone areas and causal factors.

pattern is attributed to farmers burning agricultural waste in their fields.

Similarly, in figure 4.2f, we observe near-real-time fire alerts in the SimplyFire application. The depicted area encompasses the vicinity of the Serra do Pardo National Park in Brazil, bordering the Amazônia National Park. The verdant forested regions are depicted in a deep green hue, while areas subjected to deforestation are portrayed in a light brown color. Fire alerts from various satellites are represented as fire symbols in distinct colors. It is evident that the deforested lands are saturated with fire signatures, while there are minimal or no fire signatures in the forested areas. This pattern has emerged due to the expansion of agriculture through the clearing of forested land through deforestation, followed by the subsequent burning of the cleared areas.

If there have not been substantial alterations in such human activities, we can anticipate that wildfire signatures will adhere to historical patterns.

At a more granular level, we may also wish to analyze other parameters, such as meteorological data, including temperature, humidity, wind speed, and precipitation; vegetation data, including the types of plants, their density, and moisture content; and topographical data, detailing slope, aspect, and elevation, for a better understanding of how wildfires will progress once they are initiated. Readings of these parameters may be had using remote sensing satellites, weather stations, sensor systems, or even community observations. At this scale, we have shifted from basic fire maps towards fire simulations. We shall discuss these in depth in chapter 9 section 9.3.

4.5 DISSEMINATION OF INFORMATION

Once wildfire-prone areas have been identified or a wildfire has been detected, the subsequent action is to disseminate this information, critical data, and guidance to all pertinent stakeholders. This process is paramount for enhancing community preparedness, informing decision-making, and coordinating response efforts. Effective dissemination involves utilizing diverse communication channels [Figure 4.3] and methods tailored to different audiences, ensuring that the information is both accessible and actionable.

Wildfire portals generally disseminate information through email, SMS alerts, and WhatsApp messages [Figure 4.3a]. Social media has become a crucial platform for disseminating critical information, including fire alerts, personnel movements, equipment deployments, and updates on firefighting strategies. It also facilitates the sharing of multimedia content, such as photographs and videos. Additionally, the formation of groups enables interested stakeholders, including new staff members, citizens, journalists, and others, to easily join relevant groups and receive timely information.

For larger, more permanent datasets such as fire maps, fire compendia, etc., websites are good options [Figure 4.3b]. These resources are readily accessible to mem-

(a) Social media including WhatsApp groups can quickly transmit wildfire-management information such as detected fire points to stakeholders in the group.

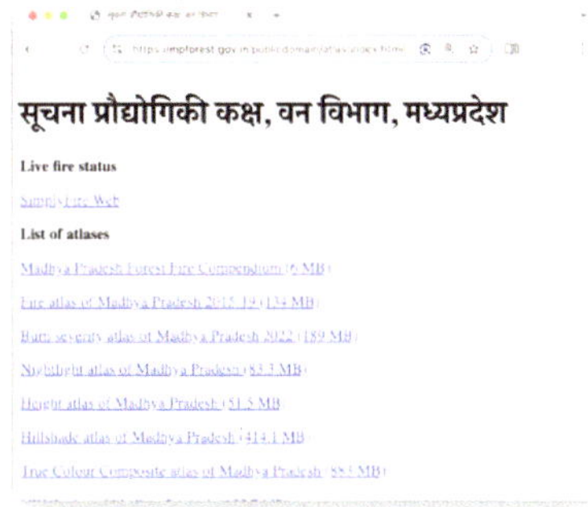

(b) Websites such as that of the Madhya Pradesh Forest Department have links to fire-management information, including real-time data on SimplyFire, fire maps, fire compendia, etc.

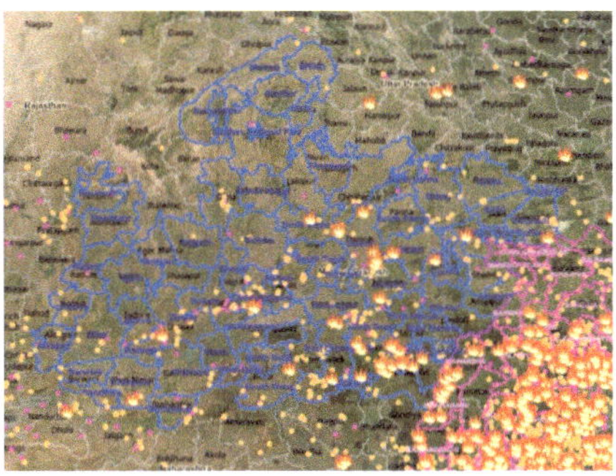

(c) Web applications such as SimplyFire display current fires at various scales. The image depicts fire alerts in the states of Madhya Pradesh and Chhattisgarh in India.

(d) Upon zooming in, SimplyFire displays fire alerts in a coal mine.

FIGURE 4.3: There are several methods to disseminate fire-related information.

bers of the public, staff, researchers, and other interested individuals. They also serve as a platform to host essential materials such as standard operating procedures and films that educate the general public about wildfire risks and preventive measures. These resources are crucial for fostering community engagement and preparedness. Furthermore, they can be utilized as reference materials to support public awareness campaigns, workshops, and community interactions.

Information pertaining to fire management professionals may be disseminated in the form of comprehensive technical reports and concise briefings that provide in-depth analyses of wildfire risks, predictive models, and strategic recommendations. These documents can be disseminated through formal channels such as agency meetings, professional conferences, and specialized publications, and may also be hosted on websites. Additionally, scenario-based training programs and simulation exercises are crucial for equipping fire management professionals with the necessary skills and knowledge to effectively manage wildfire situations. These programs assist professionals in refining their response strategies and enhancing their preparedness for actual-world incidents.

Real-time operational dashboards [Figure 4.3c] provide fire managers with a comprehensive view of current conditions, enabling them to make informed decisions regarding resource allocation and response strategies. The use of these tools enhances situational awareness and supports effective management during wildfire events. Web applications such as SimplyFire display current fires at various scales, from sub-county [Figure 4.3d] to state [Figure 4.3c], country, and global levels by integrating near-real-time satellite alerts with GIS platforms.

Coordination with local governments and agencies is facilitated through interagency meetings involving local, state, and federal agencies. These meetings enable the exchange of information and the coordination of wildfire management efforts. Collaborative databases and platforms facilitate the access and sharing of critical information, including risk assessments, resource availability, and operational plans. Urgent information regarding wildfire risks and emergency procedures is disseminated through public alert systems. These alerts, disseminated through various channels such as text messages, emails, television, radio, and automated telephone calls, provide timely updates on fire conditions and evacuation orders. Effective public alerts ensure that residents receive timely and crucial information, enabling them to take appropriate actions to protect themselves and their property.

Collaboration with community organizations, such as neighborhood associations and environmental groups, facilitates the dissemination of wildfire prevention messages. Similarly, partnerships with private sector entities, including utility companies and businesses, contribute to wildfire prevention efforts. These partners provide valuable resources by sharing information, implementing safety protocols, and supporting community preparedness measures.

Establishing feedback mechanisms is an integral component of information dissemination. It enables stakeholders to provide input on wildfire prevention initiatives and share their observations. This feedback is invaluable for refining strategies and enhancing the effectiveness of information dissemination. Collaborating with the community and other stakeholders ensures that prevention efforts are responsive to local needs, conditions, and aspirations.

4.6 COMMUNITY INVOLVEMENT AND COOPERATION

Community involvement and cooperation are fundamental to the success of wildfire prevention strategies. Engaging local residents, organizations, and non-governmental agencies fosters a collaborative approach to managing wildfire risks and mitigating their impacts. Effective wildfire prevention requires not only the implementation of technical measures but also the active participation of the community. By working together, communities can enhance their preparedness, strengthen their resilience, and ultimately contribute to a more comprehensive wildfire management strategy.

Public education is essential for wildfire prevention (Prestemon et al., 2010). Fire agencies and community groups collaborate to organize programs that educate residents about wildfire risks and preventive measures. These programs encompass various topics, such as defensible space and vegetation management. By disseminating this information, these initiatives empower residents to take proactive steps in safeguarding their properties. The Firewise Communities program (Rains and Hubbard, 2002) exemplifies successful community engagement by encouraging residents to implement fire-safe practices around their homes. Activities include creating defensible space and using fire-resistant materials, enhancing safety and fostering collective wildfire risk reduction in neighborhoods.

Developing neighborhood fire plans involves collaboration among local residents, fire departments, and organizations (Shiralipour et al., 2006). These plans outline specific strategies for preventing wildfires and executing evacuation procedures, fostering a sense of shared responsibility and preparedness. Volunteer fire departments and services [Figure 4.4] also play a crucial role in wildfire prevention, as dedicated volunteers assist with fire safety inspections and emergency response, enhancing local firefighting capabilities and supporting broader management initiatives (Lozier, 1976; Perkins, 1987).

Effective wildfire prevention necessitates robust collaboration between local governments and fire management agencies. Partnerships facilitate the exchange of resources and coordinated risk assessments, ensuring a comprehensive approach to wildfire management. A dialogue is often needed between government officials and community members [Figure 4.5], where the needs and aspirations of wildfire management are discussed, together with the requirements of the community to achieve these aspirations.

Conducting fire drills and evacuation exercises is essential for preparing communities for emergencies, allowing residents to practice procedures and identify potential challenges (Stephens et al., 2023). Regular participation in these activities enhances overall readiness for actual wildfire events. Additionally, emergency alert systems are crucial for timely communication during wildfire incidents, providing residents with updates on fire conditions and evacuation orders (Doermann et al., 2021). Community involvement in setting up these systems ensures their effectiveness and enhances residents' preparedness and responsiveness.

Promoting sustainable land use practices is also important for wildfire prevention, as responsible vegetation management and fire-resistant landscaping contribute to reducing fire hazards (Gibbons et al., 2012; Steelman and Burke, 2007). Community advocacy and education on these practices can help create safer environments, supporting long-term fire prevention efforts and enhancing community resilience.

FIGURE 4.4: Volunteer Fire Services play an important role in fire control in South Africa.

4.7 USE OF FIRE LINES AND FIRE BREAKERS

Fire lines are pivotal tools in the strategic management of wildfires, serving as both a preventative and defensive measure to control and mitigate the spread of fires (Plucinski, 2019b). Their primary role is to establish barriers that interrupt the continuity of fuel sources, thereby containing wildfires within defined limits. They are theoretically derived from fire triangles [Figure 1.3]. Since fire necessitates a combination of fuel, oxygen, and heat, the removal of fuel extinguishes a fire. Fire lines employ this principle to extinguish fires by establishing zones devoid of any vegetation matter that could serve as fuel for the advancing fire front. Comprehending the design, implementation, and maintenance of fire lines is paramount for effective wildfire prevention and management. This ensures that resources are efficiently allocated and that communities and natural resources are safeguarded from the catastrophic consequences of uncontrolled fires.

4.7.1 PURPOSE AND FUNCTION OF FIRE LINES

The primary function of fire lines is to contain and control the spread of wildfires by creating fuel-free zones that act as barriers to the advancing flames (Lemons et al., 2023). By eliminating or clearing vegetation and combustible materials from a designated area, fire lines establish boundaries that restrict the spread of a fire. This containment strategy is crucial in localizing the fire, enabling firefighting teams to

FIGURE 4.5: Dialogue between government and community members, here done as a side-event during medical camps, goes a long way in boosting cooperation.

effectively manage its containment. Fire lines also serve as a barrier, preventing fires from encroaching on unburned areas and minimizing the overall impact of the wildfire. This reduces the need for excessive firefighting resources and facilitates efficient resource allocation.

Fire lines also play a crucial role in safeguarding valuable assets, such as residential areas, infrastructure, and natural resources. By establishing fire lines around high-risk zones, communities can effectively protect homes, critical infrastructure, and sensitive ecosystems from the encroaching flames. This protective measure is particularly important in regions where wildfires pose significant threats to human safety and environmental well-being. Effective fire lines ensure that these critical assets are shielded from potential damage, thereby reducing the overall impact of the wildfire on the community and its resources. In this capacity, fire lines serve as defensive structures.

4.7.2 TYPES OF FIRE LINES

Fire lines can be categorized based on their construction method, each designed to suit specific environmental conditions and fire behaviors. The primary types of fire lines are:

- **Dozer lines**: These fire lines are created using bulldozers or other heavy machinery to clear a wide strip of vegetation and soil. Dozer lines are highly effective in combating large-scale wildfires due to their capacity to establish substantial and enduring firebreaks. They are particularly valuable in regions with high fuel loads, where they are frequently utilized in conjunction with other fire suppression techniques.
- **Hand lines**: Constructed manually by firefighting personnel using hand tools such as shovels, rakes, and hoes, hand lines are suited for smaller or more intricate areas where dozer lines may be impractical. Although less extensive than dozer lines, hand lines are valuable for creating precise firebreaks and are frequently employed in more localized or detailed fire suppression endeavors.
- **Burned-out lines**: This technique involves setting controlled burns along an existing fire line to consume any remaining fuel between the fire line and the advancing fire. Burned-out lines serve as a valuable reinforcement for containment lines by ensuring that no combustible materials remain, thereby preventing the fire from sustaining itself. This technique is frequently employed in conjunction with other fire suppression strategies to augment the effectiveness of fire lines.

Besides these, **green fire lines** may be constructed by planting of trees and vegetation that are fire-resistant and hold lots of water in their bodies; this water can then quench the fire front by removing heat through the process of evaporation. Such fire lines need to be planned quite in advance, and are not currently in large-scale operational use.

At the time of wildfire, **wet fire lines** may also be constructed by dumping large quantities of water (often in conjunction with fire retardants) in a curved line to encircle and contain the fire.

4.7.3 SITE SELECTION AND PREPARATION

Effective fire line construction requires a careful consideration of site conditions to ensure optimal performance:

- **Terrain and vegetation**: The type of terrain and vegetation influences the choice of construction method and the effectiveness of the fire line. Steep slopes, rocky terrain, and dense vegetation necessitate specialized techniques and equipment to effectively establish fire lines.
- **Weather conditions**: Weather factors, such as wind direction and speed, play a critical role in fire line construction and effectiveness. Fire lines should be strategically placed to account for prevailing wind patterns, which can affect the direction and intensity of the fire.
- **Accessibility**: Ensuring that the fire line site is accessible to firefighting

equipment and personnel is essential for both construction and maintenance. Accessibility considerations encompass the transportation of machinery, tools, and personnel to the site, as well as the ease of managing and reinforcing the fire line. Fire lines also serve as roads for the movement of vehicles, machinery, and personnel, and their width is frequently increased to facilitate their passage without impediment.

4.7.4 IMPLEMENTATION OF FIRE LINES

Construction of fire lines involves several key techniques to ensure their durability and effectiveness:

- **Clearing vegetation**: The removal of vegetation and combustible materials is fundamental to creating a fire line. This process entails the felling of trees, the removal of shrubs, and the clearance of debris to establish a clear, fuel-free barrier. Proper clearing is paramount to ensure the effective functioning of the fire line in preventing the spread of fire.
- **Grading and scarification**: In some cases, grading or scarifying the soil may be necessary to remove additional fuel material and create a more effective barrier. This technique entails breaking up the surface soil to enhance the fire line's capacity to withstand the fire's heat and prevent its advance. The process involves burying grass and debris beneath a layer of non-combustible soil. Grading and scarification are also employed to channel rainwater in a manner that does not alter the fire line, ensuring its long-term functionality over several years.
- **Reinforcement**: Fire lines may require reinforcement to maintain their effectiveness over time. This can include adding additional fuel breaks[1], reinforcing the edges with barriers and fire breakers, and conducting regular inspections to address any weaknesses. Reinforcement ensures that the fire line remains functional and effective throughout the duration of the fire.
- **Monitoring and adjustment**: Once constructed, fire lines must be continuously monitored and maintained to ensure their ongoing effectiveness. Regular inspections facilitate the identification and resolution of potential issues, including erosion, vegetation regrowth, and fire damage. Adjustments and repairs may be required in response to evolving fire conditions, weather patterns, and the wildfire's progression. Ongoing maintenance is paramount to ensure the fire line's continued effectiveness.

Effective fire line implementation necessitates collaboration among firefighting agencies, local communities, and other stakeholders. Engaging the community in comprehending the purpose and significance of fire lines, as well as involving

[1]Areas with reduced vegetation and fuel.

them in fire prevention initiatives, enhances the overall success of wildfire management strategies. Agency involvement ensures that fire lines are meticulously constructed and maintained in accordance with established best practices and that they are seamlessly integrated into comprehensive fire management plans.

Firelines are an indispensable element of a comprehensive wildfire management strategy. They are employed in conjunction with other fire suppression tactics, including aerial firefighting, controlled burns, and water drops. By integrating firelines with these methods, firefighting efforts are optimized, and the overall effectiveness of wildfire management is significantly enhanced.

4.7.5 CASE STUDY FROM MADHYA PRADESH

The state of Madhya Pradesh, situated in the central region of India, is renowned for harboring a diverse array of flagship species, including tigers, leopards, bears, and elephants. Vast sections of the state are covered with grasslands, deciduous forests, and mixed forests that are prone to wildfires [Figure 4.6a–b]. To effectively manage wildfires, construction of fire lines commences in November, frequently collaborating with local communities. This collaborative effort not only enhances wildfire control but also generates local employment opportunities. During construction, grasses, herbs, shrubs, and small plants in the selected regions are cut and left onsite to dry for approximately two months [Figure 4.6c]. This initial phase, colloquially known as "line katai" (cutting of the fire line), involves removing the vegetative material to prepare for the subsequent burning process.

In preparation of burning, ground leaf litter is removed and collected in the center of the fire line [Figure 4.6d]. Next, the dried plant biomass is set ablaze in a controlled manner. This step, referred to as "line jalai" (burning of the fire line), creates a cleared strip devoid of flammable material [Figure 4.6e]. The fire line is burned from both sides, with the burning fronts converging in the center and extinguishing each other. This final burning ensures that the fire line is effectively established and maintains its function as a barrier against the spread of wildfires.

In Madhya Pradesh, fire lines are highly effective in preventing the spread of surface fires. The fundamental principle behind fire lines is the concept of fuel exhaustion. When a fire front encounters a fire line, it encounters a barrier of incombustible bare mineral earth. With the fuel behind the fire front having already been consumed, the fire is unable to sustain itself without a continuous supply of fuel. Consequently, it begins to abate and eventually extinguishes. Simulation experiments have demonstrated that this principle is highly effective [Figure 4.6f–h] (Awadhiya, 2017).

In this manner, fire lines possess substantial capabilities in safeguarding forest patches from fires under controlled circumstances.

(a) Trees and grasses in Kanha Tiger Reserve.

(b) The forest floor has copious leaf litter.

(c) Cutting and piling of vegetation.

(d) Raking and removal of leaf litter.

(e) Burning of available fuel material.

(f) Simulated forest floor.

(g) Movement of fire front.

(h) Fire line stops fire spread.

FIGURE 4.6: Need, construction, and working of fire lines.

4.7.6 LIMITATIONS OF TRADITIONAL FIRE LINES

The effectiveness of fire lines can easily be compromised by wind, which can carry burning leaf litter across the fire line. Another significant issue is the phenomenon of bleeding [Figure 4.7a], where the fire used during the construction of a fire line escapes into adjacent forest areas, potentially burning the very forest that the fire line was meant to protect.

Additionally, traditional fire lines are susceptible to compromise from new biomass. This biomass can enter through infiltration [Figure 4.7b–d], where dry leaves shed during the dry season, and twigs and branches from trees fall over fire lines. Infiltration can also happen when winds bring dry litter to an existing fire line. New biomass can also occur through the process of repopulation [Figure 4.7e], where new grass grows in the bare soil left by the fire line. This is especially common in burned-out fire lines, since the ash that remains acts as a fertilizer for the new flush of grasses. Repopulation can occur as early as within two to three weeks, greatly diminishing the fire line's efficacy.

Mosaicking can also occur when unburnt biomass is interspersed within the fire line due to incomplete burning or disruptions caused by air movement [Figure 4.7f]. These phenomena undermine the continuity of fire lines, and such breaches can allow fire to traverse across the fire line, leading to widespread burning of the forest.

4.7.7 CONSTRUCTION OF FIRE BREAKERS

Any break in the fire line, through new biomass or left-out biomass, can gravely affect the utility of the fire line, as can be observed through simulation experiments [Figure 4.8a–c]. Any remaining biomass can serve as a bridge to allow the passage of fire to the other side of the fire line.

A way out is the construction of fire breakers [Figure 4.8d–e] — structures surrounding fire lines that aid to stop the infiltration of new biomass, and also prevent the movement of fires if the biomass has accumulated for some reason. For the construction of fire breakers, two main prerequisites can be established: the fire breakers should not be so high as to fragment wildlife habitats, and they should be constructed from locally available materials to minimize ecological impact. Local soil can be utilized directly or supplemented with water and sand to construct firebreaks. A typical width of 4.5 inches and a top height of 9 inches is often sufficient for most forest floors, although adjustments may be necessary to accommodate specific local conditions. Subsequently, the firebreaks are sun-dried on-site for a few days.

Firebreaks are simple, ephemeral, cost-effective structures designed for fire containment. They are readily constructed and maintained throughout fire seasons. It is recommended that prismatic fire breakers be used alongside traditional fire lines [Figure 4.8f] to significantly enhance their effectiveness. Firebreaks can not only prevent bleeding by providing physical barriers against the movement of leaf litter

(a) Bleeding occurs when fires escape fire lines during construction, thereby burning the forests that they were meant to protect. This is an example from Kanha Tiger Reserve, India.

(b) Infiltration is extremely common in deciduous forests where tree shed leaves in the dry seasons to conserve moisture. These dry leaves can fall over fire lines, reducing their effectiveness.

(c) Winds can also bring dried litter to fire lines, compromising their working.

(d) With time, the accumulation of dried litter can create fuel piles that are several feet deep.

(e) Repopulation by growth of new grass in a fire line constructed three weeks before in Kanha Tiger Reserve.

(f) Mosaicking happens when patches of vegetation remain unburnt during the construction of fire lines.

FIGURE 4.7: Some limitations of traditional fire lines.

(a) Simulation of a forest floor with fire line that has been breached by some biomass.

(b) The breaching biomass can act as a bridge to allow the passage of fire to the other side of the fire line.

(c) Once a fire has been able to jump across the fire line, it can burn the forest floor, or even the forest.

(d) Fire breakers can be constructed using local materials. This picture shows a cuboidal fire breaker.

(e) Fire breakers can also be constructed in a prismatic shape.

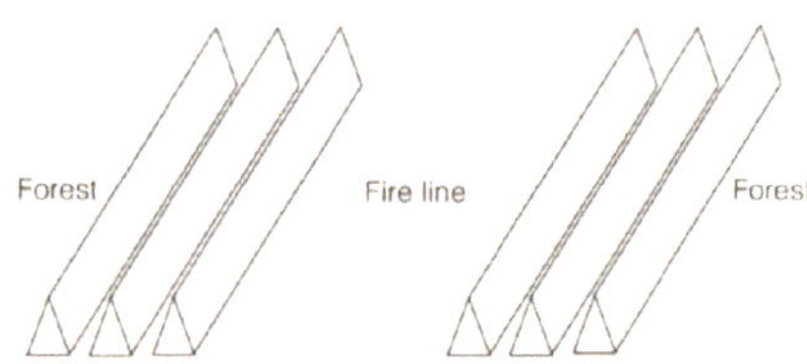

(f) Fire breakers surrounding fire lines can act as physical barriers to the movement of biomass, thus protecting the fire line from breaches. If biomass does get in, fire breakers can help prevent the movement of fire by physically restraining them.

FIGURE 4.8: The need, construction, and implementation of fire breakers.

and smoldering embers but also guard against fuel deposition through infiltration. Constructing rows of firebreaks parallel to fire lines can trap infiltration between them, which can then be regularly cleared through sweeping, blowing, or controlled burning. This approach also addresses issues of repopulation and mosaicking by allowing frequent burning of fuel loads. Firebreaks should be constructed higher than the depth of accumulated leaf litter, ideally twice to thrice the depth, while remaining low enough to avoid habitat fragmentation. Any locally available material, including soil mixed with sand or straw for reinforcement, or small pebbles and rocks, may be used.

In addition to their utilization with fire lines, fire breakers can also be deployed independently. When used alone, they should be arranged in rows to ensure a sufficient width of incombustible material for effective fire containment.

4.7.8 EFFECTIVENESS OF FIRE BREAKERS

Both cuboidal and prismatic fire breakers are effective in containing the spread of fires [Figure 4.9a–c]. When fire fronts encounter firebreaks, they cease their advance and extinguish, leaving the vegetation and dry leaves on the opposite side undamaged. Firebreaks serve as vertical barriers composed of incombustible mineral earth, effectively suppressing a fire by severing its fuel supply and preventing its passage. Their physical structure also restricts the movement of fuel and burning debris, significantly diminishing the likelihood of fire spreading. Consequently, firebreaks demonstrate superior effectiveness compared to conventional fire lines in practical field scenarios.

While simple fire lines lose their effectiveness through infiltration, cuboidal and prismatic fire breakers fare much better. Although wind-blown burning debris can occasionally re-ignite the fire on the other side while using a cuboidal fire breaker [Figure 4.9d–i], prismatic fire breakers prevent even that, since any infiltration is divided into two sections — left and right — and thus no combustible material is able to remain on top [Figure 4.9j–l]. Although a fire front can ignite the accessible flank, it cannot descend slopes to reach the second flank, thereby protecting the opposite side. Consequently, prismatic fire breakers are most effective in countering infiltration, followed by cuboidal fire breakers, while traditional fire lines are ineffective in such scenarios.

In field situations, fire breakers can not only be constructed using hand tools (providing employment opportunities to members of the local community), but also using agricultural machines such as bund makers in areas where the soil and topography permit the use of such machines.

4.8 REDUCTION OF FUEL LOAD

Reduction of fuel load is a crucial strategy in wildfire management aiming to minimize the risk and impact of wildfires (Brodie et al., 2024). Reducing the quantity

(a) Simulation of fire movement and cuboidal fire breaker.

(b) The cuboidal fire breaker stops fire movement.

(c) A prismatic fire breaker also stops fire movement.

(d) Simulation of leaf infiltration on cuboidal fire breaker.

(e) Flames are able to ignite leaves accumulated above.

(f) Top leaves may be left smoldering.

(g) Blown smoldering leaves may start fire.

(h) The started fire burns the other side.

(i) This defeats the purpose of fire breaker.

(j) A prismatic fire breaker divides leaves into two halves.

(k) Leaves on the side of the fire front catch fire.

(l) Since flames don't move down, the other side is protected.

FIGURE 4.9: Effectiveness of cuboidal and prismatic fire breakers against fire fronts and infiltration.

FIGURE 4.10: Vegetation thinning and removal of Lantana plants in Mudumalai National Park reduces fire risk while improving habitat and tourism.

of fuel available to wildfires can substantially enhance landscape resilience, thereby mitigating the likelihood of severe fires and safeguarding both human settlements and natural ecosystems. Several key practices are employed to achieve this objective:

1. **Vegetation thinning**: Vegetation thinning involves selectively removing excess vegetation, including small trees, shrubs, and underbrush. By creating more open spaces within forested areas, thinning reduces the density of flammable vegetation. This process also assists in decreasing competition for resources among plants, which can prevent the accumulation of dense, highly combustible fuel beds that contribute to intense wildfires. Thinning may also benefit certain wildlife by permitting easy movement, and also tourism since animals can be observed to larger distances [Figure 4.10]. This thinned vegetation also acts as a fuel break, and a safe place for firefighters to take stand against a wildfire.

2. **Removal of dead and decaying material**: Dead and decaying plant material, such as fallen branches and decomposed trees, can accumulate over time and become a significant fire hazard. Regular removal of this debris reduces the availability of fire-sustaining materials. This is typically achieved through manual collection, mechanical means, or by utilizing methods such as mulching to facilitate on-site degradation of the mate-

rial.

3. **Management of leaf litter and branches**: Leaf litter and fallen branches often create a continuous layer of fuel on forest floors, which can facilitate the spread of fires [Figure 4.7d]. Effective management entails the collection and disposal of this material, either through raking, shredding, or other mechanical processes. This practice helps to prevent the buildup of a substantial fuel source.

4. **Grazing and cutting**: Controlled grazing by livestock and mechanical cutting are additional methods for managing vegetation density and reducing fuel loads. Grazing can help maintain open spaces by limiting the overgrowth of grasses and shrubs. Mechanical cutting can similarly be used to manage vegetation and prevent excessive fuel accumulation (Rouet-Leduc et al., 2021).

4.9 COLD BURNING

Cold burning, also known as prescribed or controlled burning, is a technique used to manage fuel loads and reduce wildfire risk through the intentional setting of fires under controlled conditions (Miller et al., 2020). Cold burning necessitates meticulous planning, encompassing the selection of the most suitable timing contingent upon weather conditions, fuel moisture levels, and seasonal variations. Optimal conditions for cold burning are characterized by elevated humidity, diminished wind velocities, and moderate temperatures. These conditions guarantee that the fire maintains a low-intensity and controllable state.

Prior to initiating a controlled burn, firebreaks or barriers are established to confine the fire within a designated area. Continuous monitoring of weather conditions and adjustments to burning plans are essential to maintain control.

During the burn, firefighters employ techniques such as backburning and flanking to manage the fire's intensity and direction. Backburning entails igniting fuel in advance of the primary fire to establish a controlled barrier. Flanking entails burning the periphery of a fire to restrict its expansion. The primary objective is to consume low-intensity fuels while preventing the fire from escalating.

After the burn, monitoring is conducted to assess the effectiveness of the operation and to ensure that no smoldering pieces remain that may ignite a fire. This phase also includes evaluating environmental impacts and making necessary adjustments to fuel management strategies.

Cold burning serves as an effective method in reducing the accumulation of combustible materials, thereby mitigating the risk of severe wildfires. Additionally, it facilitates the growth of a new flush of vegetation that can serve as a food source for wild animals. Regular cold burning can also promote the establishment of fire-resistant vegetation through the continuous removal of fire-susceptible vegetation.

4.10 SPECIES-SELECTION FOR WILDFIRE PREVENTION

Different plant species exhibit varying influences on wildfires. Some plants shed leaves or bark that, upon drying, become readily available fuel sources. Others possess numerous thin, wispy branches that are highly susceptible to ignition. Some plants produce oily resin with high calorific values. Conversely, certain plants lack these tendencies. Succulent plants, for instance, store substantial amounts of water within their bodies, which can effectively mitigate the intensity of wildfires.

For these reasons, the selection and planting of suitable plant species can play a momentous role in wildfire prevention and management (Bond and Van Wilgen, 2012). By strategically selecting appropriate vegetation, it is possible to significantly influence fire behavior, thereby reducing the risk of wildfires and enhancing the effectiveness of other fire prevention measures. This approach entails comprehending the characteristics of various plants and how their properties can be harnessed to mitigate fire hazards.

4.10.1 CHARACTERISTICS OF FIRE-RESISTANT PLANTS

To effectively manage wildfire risk, it is essential to select plants with specific traits that make them less susceptible to ignition and more capable of reducing fire spread (Colorado and Henkin, 2023; Detweiler et al., 2006; Moore-Gough et al., 2001). The first trait is low flammability. Plants with low flammability are less likely to catch fire and contribute to the spread of wildfires. Key characteristics include:

- **High moisture content:** Plants with high moisture levels in their tissues, such as succulent leaves and stems, are less prone to ignition. Species such as the sedum (*Sedum* spp.) and various ferns exhibit high moisture content and are less flammable.
- **Low volatile oils:** Plants with low concentrations of volatile oils or resins are less likely to ignite easily. Species such as the eastern red cedar (*Juniperus virginiana*) and the white oak (*Quercus alba*) are some examples having low resin content compared to their compatriots.
- **Non-resinous substances:** Plants that produce minimal resin or gummy substances are less likely to catch fire. Examples include the California buckwheat (*Eriogonum fasciculatum*) and certain types of grass species.

We also choose plants with low crown density. Plants with a low density of foliage or branches in their upper canopy help prevent crown fires, where flames move from treetop to treetop. Trees with open, irregular crowns reduce the likelihood of fire spreading through the canopy. Non-deciduous species, such as the American beech (*Fagus grandifolia*) and the loblolly pine (*Pinus taeda*), which have less dense foliage, can be advantageous in this context.

Fire-resistant bark is another useful trait. Certain tree species with thick, fire-resistant bark can better withstand the heat of a wildfire and protect the inner tissues of the tree. Trees such as the Douglas fir (*Pseudotsuga menziesii*) and ponderosa pine (*Pinus ponderosa*) are noted for their thick bark, which provides insulation against high temperatures.

4.10.2 RECOMMENDED PLANT SPECIES

Often the best-suited plants are native species (Zouhar, 2008). Native plants are well-adapted to the local climate and soil conditions. In wildfire-prone areas, they are typically fire-resistant species that have been able to survive despite frequent wildfires. In addition, they typically require less water and maintenance and are more resistant to local pests and diseases — factors important for integrated wildfire management since dry, hollow, and perforated trees burn easily. In this way, amalgamating native species into fire management plans enhances ecological stability and resilience while concomitantly reducing fire risk.

Several plant species are particularly effective for fire management due to their fire-resistant properties:

- **Grasses:** Low-growing grasses such as buffalo grass (*Buchloe dactyloides*) and fescue varieties are effective ground covers that help reduce fuel load and limit fire spread.
- **Shrubs:** Fire-resistant shrubs like California lilac (*Ceanothus* spp.), manzanita (*Arctostaphylos* spp.), and the New Jersey tea (*Ceanothus americanus*) are beneficial plants since they have lower flammability compared to many others, while providing valuable ground cover.
- **Trees:** Fire-resistant trees such as the Douglas fir (*Pseudotsuga menziesii*), ponderosa pine (*Pinus ponderosa*), and the white oak (*Quercus alba*) are well-suited for wildfire prevention due to their thick bark and lower resin content.

Non-deciduous (evergreen) species are particularly valuable in wildfire prevention due to their all-season foliage which helps maintain ground cover while having low flammability. Examples include Ponderosa pine (*Pinus ponderosa*) (known for its thick, fire-resistant bark and ability to survive moderate fires), California bay laurel (*Umbellularia californica*) (provides dense cover but has lower flammability due to its high moisture content), and Mediterranean cypress (*Cupressus sempervirens*) (this species has a dense growth habit but acts as a barrier to fire spread due to high moisture content of its leaves).

4.10.3 LANDSCAPING PRACTICES

Landscaping practices such as creating defensible space and using firebreaks supplement the use of fire-resistant plants and accentuate their benefits (Dennis et al., 1999; Kent, 2019). Defensible space around structures involves managing vegetation to reduce fire risk. Key practices include:

- **Spacing plants:** Proper spacing between trees and shrubs reduces fuel continuity and prevents fire spread. Ensuring adequate gaps between plants may also help inhibit the movement of flames.
- **Maintaining moisture:** Regular watering of plants ensures high moisture content, reducing flammability. This practice is crucial for maintaining a low-risk environment.
- **Removing dead material:** Regular removal of dead leaves, branches, and other flammable debris prevents the accumulation of fuel that can contribute to fire spread.

Firebreaks, on the other hand, are zones devoid of vegetation or featuring low-flammability plants that act as barriers to slow or halt the progress of a fire. Incorporating fire-resistant and non-deciduous species into these areas enhances their effectiveness. Firebreaks can be constructed using fire-resistant plants or as areas left with bare soil to create a physical barrier against advancing flames, aiding the defense in depth by slowing the movement of wildfires.

5 Near-real-time detection of wildfires

Near-real-time detection of wildfires is essential for effective fire management and minimizing potential damage. This becomes crucial since with the passage of time, the fire front increases in size [Figure 3.1], necessitating the deployment of more personnel, equipment, and resources to tackle a larger fire front.

The principal technologies utilized in wildfire detection today encompass satellite-based systems, aerial surveillance, and ground-based detection methods. Satellites such as MODIS and GOES play a crucial role by providing extensive, near-real-time coverage. These satellites capture thermal anomalies and smoke plumes, offering valuable data regarding fire location, size, intensity, and movement. Aerial platforms, including drones and manned aircraft, complement satellite technology by delivering detailed, localized monitoring through thermal and infrared sensors, which are instrumental in assessing fire behavior and conditions. Ground-based manual and automated fire detection methods, including watch towers, further enhance these capabilities.

In addition to traditional methods, advancements in machine learning and artificial intelligence (AI) have significantly enhanced wildfire detection capabilities. AI algorithms analyze large volumes of data from satellite and aerial sources to accurately identify fire patterns and anomalies. Predictive modeling driven by AI further supports fire management by forecasting fire behavior based on historical data and current conditions. This also allows us to concentrate our attention and processing to areas that have a high probability of having wildfires.

Social media and citizen reports have also become integral to wildfire detection, providing real-time, crowd-sourced information. Social media platforms such as X (formerly Twitter) and Facebook, along with specialized reporting applications, provide immediate reports of wildfire sightings. These reports are utilized to corroborate and enhance data gathered from conventional detection methods.

In this chapter, we shall discuss all these in detail.

5.1 SATELLITE-BASED DETECTION

Satellite-based fire detection has become the bedrock of modern wildfire management, offering extensive, continuous, near-instantaneous, and cost-effective observational capabilities (Chuvieco et al., 2020; Wooster et al., 2021). By employing satellite technology, we can achieve comprehensive spatial coverage and frequent revisit opportunities, which are crucial for effective wildfire detection and monitoring. Satellite-based monitoring enables us to monitor remote, distant, and inaccessible areas without the need for physical human observers in those locations.

This section delves into the multifaceted role of satellite systems in wildfire detection. It elucidates the diverse data types they provide and highlights their pivotal contribution to enhancing fire management and response strategies.

5.1.1 SATELLITE SYSTEMS FOR WILDFIRE DETECTION

Several key satellite systems are employed for identifying and tracking wildfires. MODIS (Moderate Resolution Imaging Spectroradiometer), an instrument aboard NASA's Terra and Aqua satellites, provides extensive global coverage and frequent updates (Davies et al., 2008). It captures images in both visible and thermal infrared spectra, which are crucial for detecting wildfires. The thermal infrared data helps identify heat sources, while visible imagery supports tracking the fire's progression. MODIS's ability to revisit areas frequently makes it valuable for ongoing wildfire monitoring and management.

VIIRS (Visible Infrared Imaging Radiometer Suite), located on the S-NPP (Suomi National Polar-orbiting Partnership), NOAA-20, and NOAA-21 satellites, is another platform that enhances fire detection capabilities with its advanced sensors (Dalezios et al., 2017). VIIRS provides high-resolution thermal and visible images, improving our ability to detect and analyze wildfires. Its Day/Night Band (DNB) feature enables it to detect fires at all times of the day, ensuring continuous monitoring. This capability is crucial for capturing fire data during both daylight and night, making VIIRS an essential tool for comprehensive wildfire surveillance.

The GOES (Geostationary Operational Environmental Satellites) series of satellites, managed by NOAA, are positioned in geostationary orbit to provide real-time, high-frequency imagery of the same area (Chuvieco et al., 2020; Prins et al., 2001). GOES satellites offer continuous updates on wildfire activity through visible and infrared imagery. This real-time data is particularly useful for monitoring rapidly changing fire conditions and issuing timely alerts to response teams, aiding in quick and impactful deployment of resources.

5.1.2 TYPES OF DATA PROVIDED BY SATELLITES

Satellites provide diverse data types that are indispensable for detecting and managing wildfires. Thermal infrared sensors on satellites detect the heat emitted by wildfires. This data enables the precise identification of active fires, particularly in remote areas where alternative methods may be inadequate. Through the analysis of these thermal signatures, satellite data elucidates the intensity and extent of fires, which is crucial for planning firefighting strategies and resource allocation.

Similarly, optical sensors capture visible light reflected from the Earth's surface. This imagery aids in identifying smoke plumes and assessing the extent of fire-affected areas. Optical data is especially valuable for visualizing fire boundaries and comprehending the spatial spread of the fire, thereby providing insights into the extent of its environmental impact.

Certain satellites also provide multispectral or hyperspectral imagery, which captures data across a range of wavelengths. This type of data enables comprehensive analysis of vegetation health, smoke composition, and fire behavior. Multispectral and hyperspectral data significantly enhance the accuracy of fire detection and provide a more detailed understanding of wildfire dynamics.

5.1.3 APPLICATIONS IN FIRE MANAGEMENT

Satellites play a crucial role in the early detection of wildfires by identifying heat anomalies and smoke plumes. Through continuous monitoring via satellite imagery, fire progress can be tracked in real-time, facilitating prompt responses and timely updates for fire management teams. This capability is paramount in preventing the spread of wildfires and mitigating their adverse effects.

Satellites like GOES provide high-temporal resolution data that supports real-time alerts for fire agencies and responders. This immediate information is crucial for the prompt mobilization of firefighting resources and the coordination of response efforts. Real-time alerts facilitate the timely deployment of response teams to effectively manage and contain wildfires.

Through continuous data capture spanning multiple years, these satellite systems provide a consistent and comprehensive dataset that illuminates patterns and trends in fire activity. This multi-year repository of data is invaluable for developing detailed analyses, enabling researchers and policymakers to construct precise fire compendia and maps that depict the extent and frequency of fire incidents over time (Awadhiya, 2018, 2022b). Such datasets are also indispensable for comprehending the environmental and climatic factors that contribute to fire outbreaks, assessing their impact on ecosystems, and formulating strategies for wildfire management and prevention. Furthermore, the comprehensive temporal and spatial insights provided by these datasets serve as a cornerstone for modeling and predictive analyses, aiding in the anticipation of future fire behavior and mitigating risks in vulnerable regions.

Similarly, multispectral satellite data plays a pivotal role in the creation of accurate burn area maps, which are indispensable for fire management. By capturing reflective data from various wavelengths, these satellites can effectively differentiate between burned and unburned lands. This data is then processed to generate comprehensive maps that delineate the extent and intensity of the fire-affected regions (Awadhiya, 2022a). Such high-resolution maps enable fire management teams to assess damage, formulate rehabilitation strategies, allocate resources effectively, and devise preventive measures to mitigate future fires. Consequently, these efforts contribute to both ecological recovery and public safety.

5.1.4 CHALLENGES AND LIMITATIONS

Despite their benefits, satellite-based detection faces several challenges (Leblon, 2001). The resolution of satellite imagery significantly impacts the accuracy of fire detection and assessment. While higher-resolution data offers more detailed information, it may not always be feasible or cost-effective. Conversely, lower resolution can hinder the detection of smaller fires or the evaluation of fire behavior in specific environments.

Cloud cover and atmospheric conditions can impede satellite observations, thereby diminishing the capacity to identify and monitor wildfires. Despite the utilization of sophisticated algorithms and data fusion methodologies, these techniques do not completely mitigate the adverse effects of inclement weather on satellite data.

The large volume of data generated by satellites requires effective processing and interpretation. Advanced algorithms and machine learning techniques are often employed to analyze this data, but the quality of interpretation nevertheless is dependent on the data's clarity and the processing system's capabilities. Accurate interpretation is paramount for making well-informed decisions in wildfire management.

5.2 AERIAL SURVEILLANCE

Aerial surveillance plays a pivotal role in the timely detection and effective management of wildfires. By employing a range of aerial platforms — including manned aircraft, unmanned aerial vehicles (drones), and balloons — fire management teams can gain detailed and localized insights into fire activity. This section delves into the diverse aerial platforms employed for wildfire detection, elucidating the data they furnish and their pivotal role in augmenting fire management and response protocols.

5.2.1 AERIAL PLATFORMS FOR WILDFIRE DETECTION

Aerial platforms offer unique benefits for detecting and monitoring wildfires (Allison et al., 2016). Each aerial platform possesses unique advantages that collectively contribute to effective wildfire management.

Manned aircraft, particularly reconnaissance planes and water bombers, play a pivotal role in wildfire detection and management. These aircraft are equipped with sophisticated imaging systems and sensors that provide invaluable data for comprehending wildfire dynamics and devising effective responses.

Specifically engineered for meticulous observation and reconnaissance, these aerial vehicles are equipped with infrared cameras and specialized sensors designed to discern heat signatures and monitor the behavior of wildfires. The high-resolution imagery captured by these sensors enables the assessment of the intensity, extent, and impact of wildfires, providing invaluable data for the formulation of effective firefighting strategies and the efficient allocation of resources.

Primarily utilized for firefighting operations, water bombers also serve a valuable role in surveillance. Their ability to release substantial quantities of water or fire retardant on active fires is well-known. However, their capabilities extend beyond suppression efforts, enabling concurrent real-time aerial observations. This dual functionality significantly enhances their utility in wildfire management, facilitating adjustments to firefighting strategies based on evolving fire conditions.

Drones, or unmanned aerial vehicles (UAVs), have become increasingly important in wildfire detection due to their flexibility, affordability, and advanced technology. Drones utilized in wildfire detection are frequently equipped with thermal cameras to identify heat emanating from wildfires. Additionally, drones may be equipped with multispectral or hyperspectral sensors to capture data across a broad spectrum of wavelengths, enabling comprehensive analysis of vegetation health, smoke composition, and fire behavior. Drones possess the capability to operate at lower altitudes and navigate through smoke, thereby providing unobstructed and immediate visualizations of fire activity without compromising human lives. This real-time data is paramount for identifying critical fire locations and directing response efforts effectively.

Although less common, balloon surveillance is also sometimes used, utilizing high-altitude balloons or tethered balloons. High-altitude balloons, equipped with cameras and sensors, provide extended aerial observation from a stable platform. These balloons can remain aloft for extended periods, capturing high-resolution images and monitoring vast areas. Their ability to offer a comprehensive perspective on wildfire activity is invaluable for tracking the fire's extent and progression over time.

Tethered balloons, anchored to the ground, are equipped with cameras and sensors for real-time monitoring. They provide a stable and adjustable platform for observing fire activity from various heights. Tethered balloons are particularly useful for focused surveillance, offering detailed views of specific areas of interest.

5.2.2 TYPES OF DATA PROVIDED BY AERIAL SURVEILLANCE

Aerial platforms collect various types of data that are essential for effective wildfire detection and management (Allison et al., 2016). Thermal imaging data is paramount for identifying heat sources associated with wildfires. Manned aircraft, drones, and balloons equipped with thermal sensors acquire this data, enabling the pinpointing of active fires and the assessment of their intensity. This information holds particular significance in conditions of reduced visibility, such as during nighttime or in dense smoke.

Optical imagery, captured utilizing visible light sensors, offers lucid and detailed representations of wildfires and their surrounding environments. This data type facilitates the visualization of smoke plumes, fire boundaries, and the extent of fire-impacted areas. In conjunction with thermal data, optical images provide a tangible perspective on the fire's characteristics and progression, thereby supporting comprehensive assessment and response planning.

Real-time video feeds from aerial platforms, such as drones and tethered balloons, are also highly valuable for immediate decision-making. These live feeds enable fire management teams to observe fire behavior as it evolves, facilitating timely adjustments to response strategies and enhancing overall operational coordination.

5.2.3 APPLICATIONS IN FIRE MANAGEMENT

The data collected through aerial surveillance plays a pivotal role in wildfire management, enabling comprehensive assessments of wildfire conditions. Aerial platforms provide high-resolution imagery and data, enabling fire management teams to comprehend the current state of the fire. This information is paramount for making informed decisions regarding firefighting strategies, resource allocation, and evacuation planning.

Real-time data from aerial platforms further supports tactical decision-making during wildfire incidents. Immediate insights into fire behavior, location, and intensity empower incident commanders to develop and execute strategic responses. This timely information is crucial for coordinating firefighting efforts and optimizing resource deployment.

Concurrently, aerial surveillance enhances coordination and communication among firefighting teams. Data collected by aerial platforms is shared with ground crews and other responders, ensuring that all parties have access to the most up-to-date information. This collaborative approach further improves the efficiency of wildfire response efforts and ensures effective resource utilization.

5.2.4 CHALLENGES AND LIMITATIONS

Despite their advantages, aerial surveillance systems face several challenges and limitations (Allison et al., 2016; Mohapatra and Trinh, 2022). Weather conditions, including high winds, smoke, and cloud cover, can significantly impact the efficacy of aerial surveillance. These factors can diminish visibility and compromise the quality of data collected by manned aircraft, drones, and balloons. Adverse weather conditions may also restrict flight operations, thereby affecting the timeliness and accuracy of data acquisition.

The substantial volume of data generated by aerial platforms necessitates efficient processing and integration. High-resolution imagery and real-time video feeds must be analyzed promptly to provide actionable insights. The complexity of data processing can be particularly challenging, especially during rapidly evolving wildfire scenarios. Therefore, effective data integration becomes paramount for informed decision-making and coordinated response efforts.

Operating manned aircraft, drones, and balloons also entails substantial financial outlays and logistical complexities. Manned aircraft necessitate extensive resources for maintenance, fuel, and crew support. While drones offer cost-effectiveness, they also demand meticulous management of battery life, maintenance, and operational safety. Balloons, regardless of altitude or tethering, necessitate similar considerations for equipment and deployment. Striking a balance between these factors is paramount for maintaining effective aerial surveillance capabilities.

5.3 GROUND-BASED DETECTION SYSTEMS

Ground-based detection systems play a critical role in the near-real-time detection and monitoring of wildfires (Mohapatra and Trinh, 2022). These technologies complement aerial and satellite-based methods by providing localized and detailed observations. These technologies include fire towers, sensor networks, and other specialized equipment that enhance our ability to detect and respond to wildfires effectively. This section provides a comprehensive examination of the various ground-based detection systems, their functionalities, and their contributions to wildfire management.

5.3.1 FIRE TOWERS

Historically, fire towers have been pivotal in ground-based wildfire detection. These structures provide elevated vantage points, enabling observation of vast areas for the detection of smoke and heat.

Traditional fire towers are constructed on elevated terrain, such as hilltops or ridges, providing a broad view of the surrounding landscape (Awadhiya, 2018; Berryoung, 2021). These fire towers are staffed by fire lookout personnel who diligently monitor the horizon for any indications of smoke or fire. Lookout personnel utilize telescopes and other optical instruments to visually scan for smoke plumes or flames. This method enables the identification of potential wildfire locations and initial assessments of fire size and behavior. Upon detecting a fire, personnel in fire towers manually report the sighting to fire control centers. This information is subsequently utilized to coordinate further response actions and dispatch additional resources as deemed necessary.

Modern fire towers have evolved to include advanced technologies that enhance their detection capabilities (Çetin et al., 2013). Contemporary fire towers are frequently equipped with automated detection systems that utilize image analysis to discern smoke and heat signatures. These systems employ algorithms to analyze camera feeds and identify potential fires with exceptional accuracy. Automated alerts are promptly generated upon fire detection, facilitating prompt response. Furthermore, these advanced fire towers are integrated with communication networks that facilitate real-time data transmission to fire management agencies. This integration enhances coordination and ensures the availability of accurate and timely information for decision-making purposes.

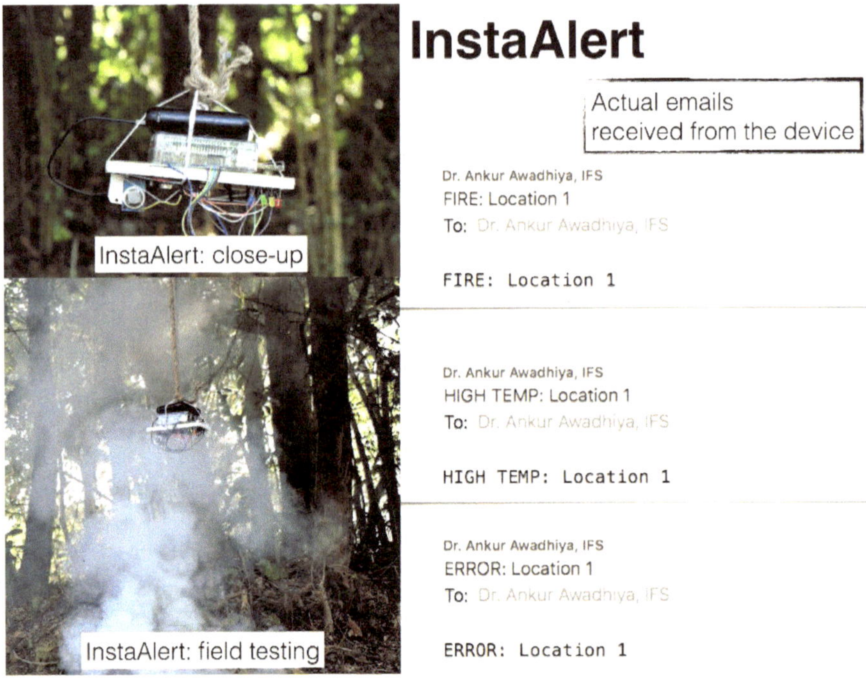

FIGURE 5.1: InstaAlert — an example of a fire detection sensor. Fire alerts are sent when smoke is detected, high temperature alerts are sent during hot periods to indicate a high propensity of fire, and error alerts are sent if any sensor goes offline.

5.3.2 SENSOR NETWORKS

Sensor networks [Figure 5.1] are distributed systems designed to monitor environmental conditions and detect signs of wildfire activity (Dogra et al., 2021). These networks consist of various types of sensors placed across a landscape to gather data continuously (Allison et al., 2016; Rahman et al., 2021; Vejmelka et al., 2016).

Environmental sensors monitor pivotal variables that influence wildfire risk and detection. Temperature sensors measure ground and air temperatures, offering data on temperature fluctuations that may suggest the presence or propensity of a fire. Sudden temperature surges can prompt alerts for potential wildfire activity.

Humidity sensors monitor fluctuations in the moisture content of the air and soil. Low humidity levels can increase the risk of wildfires, and abrupt drops in moisture may indicate the presence of a fire or a heightened fire danger.

Wind sensors measure wind speed and direction, which are crucial for comprehending fire behavior and propagation. Elevated winds can intensify wildfire conditions, and monitoring wind patterns facilitates the prediction of fire movement and potential consequences.

Ground-based smoke detectors are meticulously engineered to discern the presence of smoke particles suspended in the air. Optical smoke detectors employ laser, ultraviolet, or infrared technology to accomplish this task. Through the analysis of alterations in light intensity or wavelength, optical smoke detectors can effectively identify smoke even at low concentrations.

Chemical smoke detectors analyze the composition of airborne particles to detect the presence of smoke and other combustion byproducts. These detectors can provide additional information regarding the fire's intensity and the type of fuel being burned.

Ionization smoke detectors utilizing radioactive sources like americium-241 (^{241}Am) are generally avoided in wild areas due to radiation issues if they get damaged or lost in the environment.

5.3.3 ADVANTAGES OF GROUND-BASED DETECTION SYSTEMS

Ground-based detection systems provide several advantages in wildfire detection and management. They offer localized and detailed observations that complement aerial and satellite-based methods. By focusing on specific areas, these systems can detect smaller fires or early signs of fire activity that might be missed by other methods.

Numerous ground-based systems, including sensor networks, provide continuous monitoring of environmental conditions with exceptional detail. This is unlike geostationary satellites that provide coarse spatial resolution data owing to their humongous distance from the Earth's surface, and also unlike the sun-synchronous orbit Earth-observation satellites that have low temporal resolution, and thus cannot provide continuous data. Continuous ground-based surveillance enables the identification of potential changes that may indicate the presence of a wildfire, facilitating prompt detection and response.

Ground-based detection systems are frequently more cost-effective than aerial or satellite-based methods. Although they may necessitate initial investment and maintenance, the ongoing operational costs are typically lower. This cost-effectiveness renders ground-based systems a practical choice for extensive and long-term wildfire monitoring.

5.3.4 CHALLENGES AND LIMITATIONS

Despite their advantages, ground-based detection systems encounter several challenges and limitations. Ground-based systems, particularly fire towers, are constrained by their geographical location and the range of their observations. Their effectiveness is limited to the areas within their line of sight or sensor network coverage. This limitation can be mitigated by deploying multiple systems, but this may result in an increase in overall costs.

Ground-based systems also necessitate regular maintenance to ensure optimal functionality. Fire towers and sensors must undergo periodic inspections and repairs as required, which can be particularly challenging in remote or inaccessible locations. Furthermore, harsh environmental conditions can compromise the durability and reliability of equipment.

Although situated closer to the action, ground-based detection systems may sometimes experience longer response times compared to real-time aerial or near-real-time satellite-based methods. While automated systems can promptly generate alerts, the manual observation and reporting of fires from fire towers can introduce delays. Efficient communication and coordination are crucial to mitigate this issue.

5.4 SOCIAL MEDIA AND CITIZEN REPORTS

Social media and citizen reports have become increasingly valuable in the near-real-time detection and monitoring of wildfires (Slavkovikj et al., 2014). As wildfires can spread rapidly and impact diverse regions, integrating information from the public and social media platforms can augment conventional detection methods and optimize wildfire response strategies. This section delves into the role of social media and citizen reports in wildfire detection, elucidating their advantages, methodologies, and challenges.

5.4.1 ROLE OF SOCIAL MEDIA IN WILDFIRE DETECTION

Social media platforms such as X (formerly Twitter), Facebook, Instagram, and others provide a wealth of real-time information that can be utilized for wildfire detection (Lei et al., 2018; Slavkovikj et al., 2014).

Social media monitoring entails analyzing posts, images, and other content shared by users to discern indicators of wildfire activity. This approach leverages social media platforms to provide immediate updates on wildfire sightings, conditions, and impacts. Citizens frequently contribute to wildfire awareness by posting about fires they observe, sharing photographs or videos, and providing descriptions of fire activity. This real-time information facilitates the identification of novel wildfire incidents and enhances situational awareness for fire management teams.

Automated tools and algorithms are frequently utilized due to their ability to process substantial volumes of social media data, thereby facilitating the identification of pertinent posts pertaining to wildfires. Natural language processing (NLP) techniques analyze text for keywords related to fires, while image recognition algorithms assess visual content for signs of smoke or flames (Khurana et al., 2023; Ma et al., 2024). This automated analysis facilitates the filtering and prioritization of information for further investigation.

Many social media posts include geotags, which provide location information about where the content was created and/or shared (Crampton et al., 2013). Geotagged posts provide precise location data that can aid in pinpointing the origin and spread of wildfires. This geographic information enhances the capacity to monitor fire activity and evaluate its impact on specific regions.

Social media data can also be used to create crowdsourced maps of wildfire activity (Bogdos and Manolakos, 2019; Zhong et al., 2016). These maps integrate data from multiple sources, including user-generated content and official reports, to offer a comprehensive overview of fire locations and impacted regions. Crowdsourced maps are highly valuable for situational awareness and coordinating response efforts.

5.4.2 ROLE OF CITIZEN REPORTS IN WILDFIRE DETECTION

Citizen reports provide a direct and immediate source of information regarding wildfire activity, as reported by individuals on the ground. Various mobile applications and online platforms allow citizens to report wildfire sightings and conditions directly to authorities (Athanasis et al., 2015; Monedero et al., 2019; Nayebi et al., 2017).

Reporting applications are typically designed with intuitive interfaces that facilitate user submission of information regarding wildfires. These interfaces often incorporate functionalities for clicking and/or uploading photographs, providing detailed descriptions, and marking locations on a map. The user-friendly nature of these applications encourages widespread participation, thereby providing valuable data for fire management agencies. Furthermore, numerous reporting applications are integrated with emergency management systems, facilitating the direct transmission of reports to fire control centers. This integration expedites verification and response to reported incidents, thereby enhancing overall wildfire detection and management capabilities.

Citizen reports frequently rely on community involvement and local knowledge. This is advantageous because community members often possess comprehensive knowledge of their surroundings and can provide insights into fire activity that may not be detected by other detection methods. Local observations are particularly valuable in remote or rural areas where conventional detection systems may be less effective. Citizen reports are often verified through follow-up by fire management teams or through cross-referencing with other data sources. Validating citizen reports ensures the accuracy of information and helps prevent false alarms.

5.4.3 BENEFITS OF SOCIAL MEDIA AND CITIZEN REPORTS

Leveraging social media and citizen reports presents several advantages for wildfire detection and response. Social media and citizen reports provide real-time, on-the-ground perspectives of wildfire activity, augmenting situational awareness. This information enables fire management teams to comprehend the current state of fires, identify emerging threats, and modify response strategies accordingly.

Social media and citizen reports augment the coverage of wildfire detection beyond conventional methods. By pooling the collective observations of individuals from various locations, fire management agencies can identify fires that might otherwise remain undetected, especially in regions with sparse sensor coverage or remote locations.

The immediacy of social media and citizen reports facilitates prompt detection and response to wildfire incidents. Real-time updates and direct reports empower fire management teams to act expeditiously, mobilizing resources and implementing response measures in a timely manner.

5.4.4 CHALLENGES AND LIMITATIONS

Despite their advantages, social media and citizen reports encounter several challenges and limitations in wildfire detection. The sheer volume of data generated by social media and citizen reports can be overwhelming. Effectively managing and filtering this information to identify relevant and accurate reports requires robust data management and filtering systems. Information overload can result in delays in response and difficulties in prioritizing reports.

Simultaneously, the veracity and dependability of social media and citizen reports can fluctuate. Posts and reports may contain inaccuracies, misinformation, or exaggerated assertions. Verification procedures are crucial to establish the credibility of reports, thereby preventing false alarms and ensuring the utilization of accurate information for decision-making purposes.

The use of social media and citizen reports also raises privacy concerns and data management issues (Barbier et al., 2012; Li et al., 2016, 2017). Collecting and analyzing data from individuals necessitates meticulous handling to safeguard privacy and adhere to data protection regulations. Striking a balance between the necessity for information and privacy considerations is a crucial aspect of effectively utilizing social media and citizen reports.

5.4.5 INTEGRATION WITH OTHER DETECTION METHODS

Social media and citizen reports are most effective when integrated with traditional wildfire detection methods. This data fusion involves combining social media and citizen report data with information from satellite, aerial, and ground-based systems to provide a comprehensive view of wildfire activity. Data fusion techniques integrate diverse datasets to enhance detection accuracy and improve overall situational awareness.

In conjunction with official detection methods, social media and citizen reports can be utilized to coordinate response efforts. By integrating public information into emergency management systems, fire agencies can enhance their planning and execution of response strategies, thereby achieving a more effective and coordinated approach to wildfire management.

5.5 INTEGRATED DETECTION SYSTEMS

Integrated detection systems represent a sophisticated approach to wildfire monitoring and management, combining multiple technologies and data sources to enhance detection capabilities and response efficiency (Ambrosia et al., 1998). By integrating diverse methodologies, including satellite imagery, aerial surveillance, ground-based sensors, machine learning, and social media, these systems offer a more comprehensive perspective on wildfire activity. This section delves into the advantages and disadvantages of integrated detection systems, as well as their practical applications.

5.5.1 COMPONENTS OF INTEGRATED DETECTION SYSTEMS

Integrated detection systems combine a diverse array of technologies and data sources into a comprehensive framework for wildfire detection. Satellite and aerial imagery are pivotal components, providing both extensive and granular coverage of wildfire activity. Satellites offer continuous, global monitoring capabilities through their thermal and optical sensors, capturing extensive fire data and tracking its evolution over time. Aerial surveillance, utilizing drones and manned aircraft, delivers high-resolution, localized observations, enabling detailed insights into fire behavior and facilitating targeted assessments of specific regions.

Ground-based sensors are another crucial component, encompassing fire towers and distributed sensor networks. Fire towers, equipped with optical and thermal cameras, provide elevated vantage points for detecting smoke and heat. In contrast, sensor networks monitor environmental parameters such as temperature, humidity, and wind speed. This real-time data enables the assessment of fire danger and the early identification of fire activity.

Social media and citizen reports further enhance the integrated detection system. Social media platforms, such as X (formerly Twitter) and Facebook, provide immediate updates on wildfire sightings and conditions through user-generated content. Geotagged posts and images offer location-specific insights that complement satellite and aerial data. Citizen reporting applications further augment detection capabilities by enabling individuals to report wildfire sightings directly, providing valuable on-the-ground information that aids in tracking and responding to fires, particularly in less monitored or remote areas.

5.5.2 BENEFITS OF INTEGRATED DETECTION SYSTEMS

Integrated detection systems provide several advantages that enhance wildfire detection and management. By combining data from multiple sources, a comprehensive and detailed view of wildfire activity is achieved, leading to improved detection accuracy and a deeper understanding of fire behavior. This approach tends to reduce the likelihood of missed wildfires while simultaneously enhancing the confidence in the detected wildfires.

Access to real-time information from a multitude of sources also enhances situational awareness, facilitating more informed decision-making and effective response strategies.

Integrated systems frequently enhance resource management capabilities by accurately predicting fire behavior and evaluating current conditions. This enables strategic deployment of firefighting resources.

5.5.3 CHALLENGES AND LIMITATIONS

Despite their advantages, integrated detection systems face several challenges (Ambrosia et al., 1998; Lourenço et al., 2021). Integrating data from disparate sources necessitates the application of sophisticated data mining and data fusion methodologies to guarantee the accuracy and uniformity of the resulting system.

Integrating multiple technologies and data sources presents a complex challenge, necessitating advanced infrastructure and effective coordination. Simultaneously, implementing and maintaining integrated detection systems entails substantial financial outlays for technology, infrastructure, and human resources.

5.5.4 CASE STUDIES AND EXAMPLES

Examining real-world implementations highlights the effectiveness of integrated systems. California's integrated network[1] combines satellite imagery, aerial surveillance, ground-based sensors, and social media monitoring to enhance wildfire detection and response.

European countries similarly use integrated systems incorporating satellite data, ground-based sensors, and predictive modeling for cross-border fire management (Kutchartt et al., 2023).

India has developed SimplyFire[2] [Figure 5.2], a web-based application that integrates various satellite and topographic data. This capability enables the timely detection of wildfires, facilitating the development of efficient and effective fire control strategies by providing access to the three-dimensional topography of the affected area.

[1]https://hub.wftiic.ca.gov

[2]https://mpforest.gov.in/publicdomain/atlas/Index.html

FIGURE 5.2: Near-real time wildfire alerts in the Himalayas seen on SimplyFire 3D web application. Information about local topography aids planning and preparation for fire-fighting operations.

6 Wildfire suppression

Wildfire suppression involves a variety of techniques aimed at controlling and extinguishing fires to minimize damage and protect both natural and human environments (Plucinski, 2019a). Key methods include creating firebreaks, which are gaps in vegetation or combustible materials designed to stop or slow the spread of a fire. These barriers can be established using heavy equipment such as bulldozers and graders, which clear large areas of vegetation. Alternatively, hand tools such as shovels and rakes can be employed for more detailed work. In certain instances, explosives are also utilized to swiftly clear fuel and establish firebreaks, although their application necessitates meticulous planning to prioritize safety and efficacy.

Controlled burns and counter burning are strategic techniques used to manage wildfire risks and reduce fire intensity. Controlled burns, also known as prescribed fires, involve intentionally setting fires under controlled conditions to manage fuel loads and improve ecosystem health. Counter burning, or backing fires, involves setting a fire in front of the main blaze to consume available fuel and slow its progress. Both methods are designed to reduce the flammability of the environment and prevent the spread of large, uncontrollable fires.

Aerial suppression is another critical component of wildfire management. This method utilizes aircraft to deliver water, fire retardants, or other suppressants directly onto or around the fire. Water bombing, a specific technique within aerial suppression, involves dropping large volumes of water from aircraft to cool the fire and reduce its intensity. This method is frequently employed in conjunction with other suppression techniques to effectively manage extensive and remote wildfires. Specialized fire suppression teams and advanced monitoring technologies, such as satellites and drones, further support these efforts by providing real-time data and enabling rapid, targeted responses to wildfire threats.

In this chapter, we will discuss a comprehensive analysis of various methods employed for wildfire suppression.

6.1 PRINCIPLES OF WILDFIRE SUPPRESSION

Effective wildfire suppression is underpinned by several fundamental principles that guide strategies and tactics in controlling and extinguishing wildfires while minimizing their impact on human life, property, and ecosystems. By comprehending and applying these principles, a coordinated and effective response to wildfires can be achieved.

The first task is to contain the fire to defined limits. **Containment** is a primary goal in wildfire suppression, focusing on preventing the fire from spreading beyond established boundaries (Fried and Fried, 1996; Hu and Ntaimo, 2009). This principle involves creating and maintaining firebreaks — cleared areas devoid of combustible materials that act as barriers to the fire's progression (Ascoli et al., 2020; Plucinski, 2019b). Containment lines can be natural, such as rivers or rocky ridges, or man-made, such as trenches or cleared paths (Plucinski, 2019a). The effectiveness of containment strategies depends on the thoroughness of the firebreak and the fire's behavior, which can be influenced by wind, terrain, and fuel availability. Successful containment effectively restricts the wildfire's expansion, safeguarding unburnt regions. This measure ultimately diminishes the wildfire's overall impact.

The second task is to reduce the contained fire's intensity and spread. **Control** measures are aimed at directly reducing the fire's intensity and spread (Plucinski, 2019a). This principle involves various techniques, including the application of water and fire retardants (Underwood and Held, 2012), the execution of controlled burns (Miller et al., 2020; Wagle and Eakle, 1979), and the use of heavy equipment to manage the fire's behavior (Lueck, 2013). Water and fire retardants work by cooling the fire and creating a chemical barrier that inhibits combustion (Grant et al., 2000; Green, 1995; Hua et al., 2002; Mawhinney et al., 1994). Controlled burns, or prescribed fires, are intentionally set to reduce fuel loads and modify fire behavior in a controlled manner (Francos and Úbeda, 2021; Yoder et al., 2004). Heavy equipment, including bulldozers and graders, is employed to establish firebreaks and eliminate combustible materials. Control measures are tailored to the specific circumstances of the fire and the surrounding environment, with the objective of reducing the fire's intensity and rendering it more controllable.

Next, we aim to extinguish the fire. **Suppression** encompasses all actions taken to extinguish the fire or significantly reduce its size and intensity. This principle involves deploying specialized firefighting teams, using aerial resources such as water bombers and helicopters, and employing a range of tools and techniques to combat the fire (Kennedy, 2019; Underwood and Held, 2012). Ground-based efforts include constructing containment lines, removing vegetation, and directly attacking the fire with hand tools. Aerial suppression involves dropping water or fire retardants to quickly address large or remote areas (Plucinski and Pastor, 2013). The objective of suppression is to extinguish active flames, cool down hot spots, and eliminate potential sources of re-ignition. Effective suppression necessitates a coordinated approach that integrates various methods to address the fire's diverse aspects and stages.

While tackling fire, we need to be mindful of the safety of firefighters and the public, and the efficiency of our operations. **Safety** is a critical principle in wildfire suppression, prioritizing the protection of firefighters, emergency responders, and the public (Plucinski, 2019b; Tedim et al., 2020a). Ensuring safety involves assessing potential risks, implementing safety protocols, and providing personal protective equipment for responders. This principle encompasses ongoing risk assessments to adapt to evolving circumstances and prevent accidents. Effective communication, training, and adherence to safety guidelines are paramount for safeguarding all individuals involved in firefighting endeavors. By maintaining a steadfast commitment to safety, suppression operations can be executed more effectively, thereby minimizing the likelihood of injury or harm.

Efficiency in wildfire suppression aims to optimize the use of resources, including personnel, equipment, and materials, to achieve the best possible outcomes. This principle entails strategic planning and prioritization, ensuring that resources are allocated in accordance with the fire's behavior, the available capabilities, and the overall objectives of the suppression effort. Efficient operations minimize costs, reduce response times, and enhance the effectiveness of suppression efforts. By balancing resource availability with operational requirements, fire management can address the fire more swiftly and effectively, thereby reducing its overall impact.

Since wildfire locations are diverse, with multiple fuels, terrain, and weather conditions [Figure 6.1a–b], a variety of site-specific measures are needed to efficiently tackle and control wildfires [Figure 6.1c–d].

The availability of modern technologies, including satellites for detection of wildfires, has greatly increased the efficiency of wildfire suppression operations by providing near-real-time alerts of fire locations [Figure 6.2a]. Since wildfires increase in size with the passage of time [Figure 3.1], we need to reach the wildfire spot as soon as possible. In this regard, the availability of graphical mapping tools and applications such as SimplyFire[1] has helped increase efficiency further. Before such applications, the firefighting teams often made use of hand-held GPS devices. In such devices, the location of a wildfire can be keyed in as a waypoint, and the device helps to move in a straight line to the desired waypoint [Figure 6.2b]. In theory, this concept is advantageous because the straight-line path is also the shortest path. Consequently, traversing this shortest path, "as a crow flies," frequently yields faster travel compared to an erratic route.

However, empirical observations indicate that this direct path, while shortest in terms of distance, is frequently not the least time-consuming. This is because the path may be obstructed by dense vegetation, water bodies, or hillocks, impeding the movement of the fire-fighting team. A much better path would be through the fire lines or roads, that while longer, are relatively unobstructed, permitting faster movement of personnel and equipment [Figure 6.2c].

Adaptability is also essential in wildfire suppression due to the dynamic nature of fires and changing environmental conditions. This principle entails continuously evaluating the fire's behavior, weather conditions, and other factors that may influence suppression efforts. Adaptability necessitates flexibility in tactics and strategies, enabling adjustments based on real-time information and changing circumstances. Effective adaptation ensures that suppression efforts remain pertinent and responsive, optimizing the overall management of the wildfire.

---

[1]http://mpforest.gov.in/publicdomain/atlas/index.html

(a) In uneven terrains, wildfires can quickly spread up the slopes, but movement of personnel may be difficult. Image of wildfire in hills of Balaghat, India.

(b) Grasslands are even terrains, but availability of quick-burning fuel in large quantities makes wildfire management challenging. Image of Velavadar National Park, India.

(c) In dense or mountainous areas, helicopters allow fast movement of trackers and personnel. Image from Kruger National Park, South Africa.

(d) On even terrains, ground vehicles are used to move personnel and equipment. Image from Cape Town, South Africa.

FIGURE 6.1: Wildfires occur in different terrains and require site-specific management.

Collaboration emphasizes the importance of coordinated efforts among various agencies, organizations, and stakeholders involved in wildfire suppression. This principle emphasizes the importance of effective communication, resource sharing, and joint planning among local, state, and federal entities, as well as community members. Collaborative approaches enhance the overall response to wildfires by pooling expertise, coordinating resources, and ensuring a unified strategy. Through cooperation, diverse groups can address the complexities of wildfire management more effectively, resulting in improved outcomes and a more comprehensive response.

6.2 WILDFIRE SUPPRESSION METHODS

Wildfires can be suppressed through an understanding of the fire triangle [Figure 1.3a], which consists of three essential components: heat, fuel, and oxygen. Effec-

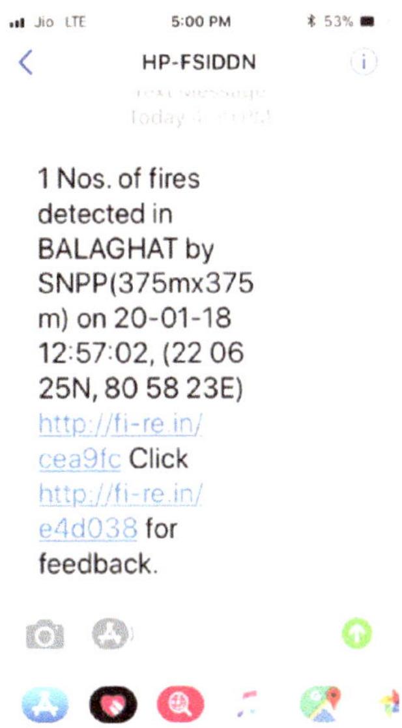

(a) SMS alert of near-real-time detection of wildfire.

(b) Movement of personnel to wildfire location in a straight line is often more time-consuming.

(c) Movement of personnel through roads and fire lines is often less time-consuming.

FIGURE 6.2: SimplyFire increases efficiency of personnel movement by suggesting movement along roads and fire lines.

tive suppression methods seek to disrupt this triangle by targeting one or more of its elements. For instance, creating firebreaks or removing vegetation eliminates the fuel source, depriving the fire of one of its essential components. The application of water or fire retardants reduces the fire's temperature and inhibits combustion, addressing the heat aspect. Fire retardants also work by creating a barrier that cuts the supply of air or oxygen. By systematically disrupting these critical elements, suppression strategies effectively diminish the fire's capacity to sustain itself, leading to successful containment and extinguishment. Let us delve into some of the wildfire suppression methods in greater detail.

6.2.1 FIREBREAKS

Firebreaks are an essential element of wildfire management, playing a crucial role in both wildfire prevention and suppression (Ascoli et al., 2020; Plucinski, 2019b). These barriers, created by clearing gaps in vegetation or other combustible materials, effectively slow down or halt the advance of a wildfire. By establishing these breaks, we safeguard valuable resources, infrastructure, and communities while simultaneously supporting firefighting efforts.

Firebreaks can be either natural or man-made. Natural firebreaks encompass geographical features that serve as robust barriers to wildfires. Rivers, lakes, cliffs, and rock outcrops are notable examples of natural firebreaks, providing substantial obstacles that flames cannot readily overcome. Furthermore, wetlands such as marshes or bogs can function as firebreaks due to their high moisture content, which renders vegetation less flammable and impedes the spread of fire.

Man-made firebreaks are intentionally constructed to effectively manage and extinguish wildfires. These firebreaks encompass cleared areas, where vegetation is systematically removed to expose bare ground or low-vegetation zones. Prominent examples of firebreaks include gravel roads, plowed fields, and cleared strips. Another method involves controlled burns, where a fire is intentionally set to burn away fuel and create a gap in the wildfire's path. Known as a "backburn", "counter burn", or "burnout," this technique is used to manage the spread of the fire by removing potential fuel sources. Physical barriers, such as fences or walls, can also be constructed to confine or decelerate the spread of a fire. Although less frequent, these structures can be effective in specific situations.

During wildfire suppression efforts, we typically strive to establish a comprehensive network of firebreaks, which may include both natural and man-made firebreaks. This harmonious network not only effectively contains the wildfire but also provides access points for further extinguishing operations. During wildfire suppression, it is often necessary to construct new firebreaks at a safe distance from the actual wildfire front to protect personnel and machinery. Additionally, counter burn operations are often conducted to widen the firebreak by consuming the combustible materials between the wildfire front and the firebreak.

6.2.1.1 Construction of firebreaks

The construction of effective firebreaks for wildfire suppression necessitates strategic planning and meticulous design. The selection of firebreak locations is contingent upon the anticipated trajectory of the wildfire, the availability of fuel, and topographical features. Ideally, firebreaks should intersect the projected path of the fire to maximize their effectiveness. The width of firebreaks typically ranges from 10 to 100 feet, influenced by factors such as fuel type, fire intensity, and the necessity for personnel and machinery movement. Wider firebreaks provide enhanced protection but necessitate substantial resources for their creation and maintenance.

Various methods are employed in the construction of firebreaks. Manual clearing involves the utilization of tools such as chainsaws, brush cutters, and hand tools to remove vegetation and combustible materials. In areas characterized by dense vegetation or challenging terrain, mechanical methods are employed, with heavy machinery like bulldozers and excavators clearing vast expanses of land swiftly. Chemical methods, such as the application of herbicides to eliminate vegetation, can also be employed, although their slow action and environmental concerns often restrict their application. In certain instances, explosives are utilized to rapidly clear vegetation or fracture rock formations. This technique proves particularly effective in challenging or inaccessible terrains where conventional methods may prove impractical.

6.2.1.2 Limitations and challenges

Although firebreaks are essential for wildfire management, they also face certain limitations and challenges. We have already discussed the effects of bleeding, infiltration, repopulation, and mosaicking [Figure 4.7]. Furthermore, the effectiveness of firebreaks can be diminished in regions with highly flammable fuels, such as oily or resinous plants. Extreme weather conditions, including high winds or intense heat, can also compromise their efficacy by dispersing embers across the firebreak, potentially facilitating the fire's spread. Furthermore, the construction and maintenance of firebreaks can disrupt local ecosystems, adversely affect wildlife habitats, and lead to soil erosion. Firebreaks can also lead to habitat fragmentation by creating breaks in ecosystems and altering animal movement patterns, thus disrupting the continuity of habitats, potentially affecting biodiversity and leading to changes in local ecosystem dynamics. The process of vegetation clearance can inadvertently introduce invasive plant species, which may alter the local environment. Furthermore, effective firebreaks often incur substantial costs in terms of construction and maintenance, necessitating the allocation of land that could otherwise be utilized more productively for tree growth.

6.2.1.3 Integration with other fire management strategies

Firebreaks are not merely passive barriers but are an integral component of a comprehensive wildfire management strategy. They are employed in conjunction with various suppression techniques, including ground firefighting crews that work alongside firebreaks to directly combat the fire. Aerial support, such as aircraft dropping water or fire retardants, is also utilized to enhance the effectiveness of firebreaks. Furthermore, firebreaks are an integral part of a broader strategy that encompasses community evacuation plans and preparedness measures, ensuring a coordinated approach to managing wildfires.

By strategically positioning firebreaks in the direction of the wildfire's movement, these barriers can serve as instrumental not only in preventing the spread of fire but also in actively suppressing it. Their strategic placement and effective management enhance their role in both controlling and extinguishing wildfires, making them a crucial tool in the fight against these destructive events.

Firebreaks also function as access paths, evacuation routes, and monitoring locations where sensors can be strategically deployed. This multifaceted role further integrates firebreaks with other wildfire suppression methods, enabling their seamless integration into the comprehensive wildfire management approach.

6.2.2 USE OF EXPLOSIVES

The utilization of explosives in wildfire suppression is a specialized technique that is occasionally employed to establish or augment firebreaks, particularly in challenging terrains where conventional methods may be less effective. This approach leverages the rapid removal of vegetation and disruption of the landscape to create barriers that assist in controlling or halting the advance of a wildfire. Additionally, the pressure waves generated by explosions can be utilized to directly impact the fire itself (McKinty, 1958).

6.2.2.1 Applications of explosives in wildfire suppression

Explosives are employed to establish or widen firebreaks in regions characterized by dense vegetation or rocky terrain, where manual or mechanical clearing presents significant challenges. By detonating explosives, substantial quantities of vegetation and debris are rapidly removed, creating a wide, clear strip of land that serves as a fire barrier. This technique is particularly advantageous in remote or inaccessible areas where heavy machinery is unable to access the site.

Beyond firebreak creation, explosives can be utilized to disrupt the landscape in a manner that impedes the fire's progression. For instance, explosives can be employed to fragment rock formations or alter terrain features, rendering it more challenging for fire to spread across the landscape. This disruption can contribute to the formation of barriers that hinder fire's movement.

An additional and less frequently recognized application of explosives in wildfire suppression involves utilizing the pressure waves generated by explosions to directly impact the fire. Upon detonation, explosives produce a shockwave that can blow away flames and embers in the immediate vicinity. This action can effectively clear small, localized areas of fire or extinguish burning materials by disrupting their combustion. Although this effect is confined to smaller, manageable sections of the fire, it can provide valuable support in controlling fire spread.

6.2.2.2 Advantages of using explosives

One of the primary advantages of utilizing explosives is their remarkable speed and efficiency in clearing vast areas. Explosives can swiftly remove vegetation and debris, establishing a firebreak significantly faster than manual or mechanical methods. This capability becomes particularly crucial in situations where time is of the utmost importance, such as during the rapid advancement of wildfires, especially in areas with human habitations.

Explosives prove invaluable in rugged or inaccessible terrains where conventional methods of clearing vegetation are impractical. In mountainous regions, dense forests, or areas with dense undergrowth, explosives can effectively create firebreaks and barriers that might otherwise be challenging or impossible to achieve.

The ability of explosives to generate pressure waves that disperse flames further enhances their utility in wildfire suppression. This direct impact on the fire can effectively extinguish small pockets of fire and prevent the dissemination of embers, thereby augmenting the overall effectiveness of wildfire suppression efforts.

6.2.2.3 Challenges and considerations

The utilization of explosives in wildfire suppression entails substantial safety hazards. Handling, transportation, and detonation of explosives necessitate specialized training and strict adherence to safety protocols. There exists a risk of accidental detonation, personnel injuries, and unintended environmental damage if not managed judiciously.

The environmental impact of explosives must also be meticulously assessed. While explosives can effectively clear vegetation, they can also cause soil erosion, disrupt wildlife habitats, and lead to long-term ecological alterations. The application of explosives should be balanced with environmental considerations to minimize adverse effects.

Successful implementation of explosives in wildfire suppression necessitates meticulous planning and coordination. This encompasses determining the optimal locations for detonation, calculating the appropriate quantity of explosive material, and ensuring the implementation of all safety measures. Coordination with firefighting teams and other emergency services is paramount to effectively integrate explosive techniques into the broader wildfire suppression strategy.

Due to the complexity of use and potential side effects, this method is often retained solely as a last resort, contingent upon the failure of alternative methods.

6.2.3 BEATING OF FIRES

Beating of fires is a traditional and effective tool in wildfire suppression, often used to control small to moderate fires and prevent them from spreading (Bibby, 1978). This fire suppression method utilizes hand tools and physical exertion to extinguish or confine the fire. The primary objective is to remove or suppress the burning materials. Ground firefighters frequently employ this technique, particularly in situations where access to the fire is restricted, and alternative methods are impractical.

6.2.3.1 Techniques and tools

The primary tools employed in extinguishing wildfires include shovels, rakes, and specialized brooms or "beater" tools designed specifically for wildfire suppression. In the absence of these tools, green leafy branches of trees can be utilized as makeshift fire beaters. These tools serve various purposes, such as smothering or removing burning materials, clearing debris, extinguishing fires, and creating firebreaks.

Fire beaters or flappers are employed to effectively smother small fires or hot spots by beating down flames and embers. Additionally, shovels are utilized to move dirt and debris, cover burning embers, and establish firebreaks by excavating combustible materials. Rakes assist in clearing the area of leaves, twigs, and other combustible debris that could potentially fuel the fire.

One of the fundamental techniques in extinguishing fires is **smothering the flames** by covering them with dirt or non-combustible materials, including the flap of the fire beater. This method reduces the oxygen supply to the fire, which is crucial for combustion. Consequently, the flames are extinguished.

Furthermore, firefighters may employ hand tools to **remove or displace combustible materials** such as leaves, branches, or logs from the fire's path. By eliminating these fuels, the spread of the fire can be controlled or mitigated. Beating can also be employed to **cool down hot spots**, which are areas where the fire continues to burn beneath the surface or in smoldering embers. This prevents re-ignition and further dissemination of wildfire.

6.2.3.2 Advantages of beating fires

Beating fires is particularly useful in areas that are difficult to access with machinery or where the fire is too small for aerial support. It allows firefighters to address fires directly on the ground, making it an adaptable and versatile tool.

The method also provides a high level of precision, enabling firefighters to target specific areas of the fire, such as hot spots or embers. It is effective for managing small fires or sections of a larger fire where other methods may be less effective.

Beating fires requires minimal equipment and is relatively low-cost compared to more advanced fire suppression techniques. The user-friendly nature of the tools and methodologies renders it an accessible option for a wide range of firefighting scenarios.

6.2.3.3 Challenges and considerations

Beating fires can be risky, physically demanding, and labor-intensive. Firefighters may be required to work extended periods, frequently in challenging environments, which can be physically and mentally taxing and may lead to a decline in their effectiveness over time.

While beating fires can be effective for small to moderate fires, it is less suitable for large or rapidly spreading wildfires. In such instances, additional suppression measures, including aerial support or the utilization of heavy machinery, are necessary.

Approaching a fire also exposes firefighters to elevated temperatures, smoke inhalation, and potentially hazardous conditions. Consequently, it is paramount to ensure the use of appropriate protective gear and comprehensive training to prioritize safety during this technique.

6.2.3.4 Integration with other fire suppression methods

Beating of fires can be done to gain time till other wildfire suppression methods become accessible. It can also be done to give finishing touches to a wildfire control operation. It can also be combined with the creation of firebreaks, controlled burns, and aerial support to manage the fire more effectively. Firefighters on the ground may use beating techniques to address immediate threats, while other methods are employed to handle larger or more complex aspects of the wildfire.

6.2.4 WATER AND FIRE RETARDANTS

Water and fire retardants are essential tools in wildfire suppression. They help control and extinguish fires through direct application or by inhibiting combustion with chemicals. Each method has specific applications, advantages, and limitations, and they are often used together to manage wildfires effectively.

6.2.4.1 Water as a suppression tool

Water is commonly used for wildfire suppression and can be applied in various ways depending on the fire's size, location, and intensity (Grant et al., 2000; Hua et al., 2002; Mawhinney et al., 1994). In **direct application**, water is sprayed directly onto flames and burning materials using fire engines, helicopters, or aircraft [Figure 6.3]. This method helps cool the fire and extinguish it by reducing the heat and soaking the burning materials.

During **indirect application**, water can be used to create moist areas around the fire to prevent it from spreading. Dampening fuels reduces the likelihood of the fire moving to new areas.

Water offers several advantages in wildfire suppression. When applied to wildfires, water rapidly converts to steam, effectively reducing the rate of burning by taking away heat energy and displacing oxygen, which disrupts the fire triangle. This makes surrounding fuel material more difficult to burn. Consequently, water is highly effective in controlling wildfires.

Water is also abundant and cost-effective compared to other suppression methods. It can be sourced from nearby rivers, lakes, or reservoirs. Moreover, when used appropriately, water has minimal environmental impact. It also helps restore moisture to soil and vegetation, aiding in post-fire recovery.

However, there are some limitations. Water is less effective in large and intense wildfires, and its use often requires combination with other methods. This is because water evaporates quickly in high temperatures, reducing its effectiveness if not applied in sufficient quantities.

Additionally, delivering water to the fire can be challenging, especially in remote or rugged terrains. This may necessitate aerial support or temporary water delivery

(a) Fire engines are often utilized for wildfire control operations.

(b) Helicopters may be employed for coordination, and for dropping of fire retardant.

FIGURE 6.3: Ground and aerial vehicles are often brought into service when using fire retardants.

systems, which can be expensive and complex to execute. Consequently, water is often accompanied by other fire retardants.

6.2.4.2 Fire retardants

Fire retardants are chemicals designed to slow down or inhibit the spread of fire. They come in various forms, including:

- **Aqueous Film-Forming Foams (AFFF):** These foams create a protective layer on surfaces, preventing oxygen from reaching the fuel and reducing flammability.
- **Dry Chemical Retardants:** Powders that are spread over the fire to suppress flames and inhibit combustion.
- **Gel-Based Retardants:** Sticky substances that adhere to surfaces, providing a longer-lasting barrier against fire.

Fire retardants have several advantages (Lu et al., 2023; Yu et al., 2019). They inhibit combustion by altering the chemical reactions involved, thereby hindering the fire's progression. This can significantly reduce the fire's spread and provide firefighters with a buffer.

Furthermore, fire retardants exhibit extended effectiveness, with many retardants, such as Phos-Chek, maintaining their potency for an extended period, much greater than the duration reached using application of water alone. This ensures continuous protection even after initial application.

Additionally, fire retardants can swiftly establish firebreaks or safeguard critical infrastructure by effectively slowing the fire's spread. This allows firefighters to manage the blaze more effectively.

However, there are also some limitations. Some fire retardants, including Phos-Chek, contain chemicals that can be harmful to the environment, affecting soil, water sources, and vegetation (Buscemi et al., 2002; Dietrich et al., 2014; Lanctôt et al., 2024; Tunstill et al., 2022). It is crucial to evaluate the environmental impact and prioritize the use of more environmentally friendly alternatives whenever feasible.

Fire retardants are typically more costly than water, necessitating meticulous planning and resource allocation. Weather conditions, particularly wind and temperature, can influence the dispersion and efficacy of retardants, potentially diminishing their performance.

6.2.4.3 In detail: Phos-Chek

One of the most widely utilized fire retardants is Phos-Chek. It is a long-term fire retardant that comprises phosphate and other chemicals to inhibit combustion. Phos-Chek retardants are available in various forms, including powders and liquid concentrates, and are designed to provide extended protection against fire.

Phos-Chek products typically consist of ammonium phosphate, ammonium sulfate, and additives. These chemicals function by forming a protective coating on the fuel, thereby suppressing flames and reducing flammability. Phos-Chek is typically applied by aircraft or ground equipment. Aircraft can disperse it over extensive areas swiftly, while ground crews apply it directly to specific fire zones or areas of concern.

Phos-Chek provides extended protection and can be effective for several days to weeks after application. It is useful for establishing firebreaks, safeguarding infrastructure, and controlling fire spread. However, the utilization of Phos-Chek, akin to other fire retardants, entails known environmental implications. Its chemical constituents can impact soil, water sources, and vegetation. Additionally, there is a substantial cost associated with its use, and its effectiveness can be influenced by weather conditions.

6.2.4.4 Integration of water and fire retardants

Water and fire retardants are frequently employed in conjunction to enhance their effectiveness. Water can be utilized for initial cooling and direct suppression, while fire retardants such as Phos-Chek can establish enduring protective barriers and effectively control fire spread. This combination facilitates a comprehensive approach to wildfire management, addressing both immediate threats and long-term control. Thus, water and fire retardants can easily be integrated with other aspects of comprehensive wildfire management.

6.2.5 AERIAL SUPPRESSION AND WATER BOMBING

Aerial suppression and water bombing are critical techniques used in wildfire management, particularly for controlling large or rapidly spreading fires (Graham, 2020; Kennedy, 2019; Plucinski and Pastor, 2013; Underwood and Held, 2012). These methods involve deploying aircraft to deliver water, fire retardants, or other chemicals directly onto the fire or its surrounding areas. They are valuable for their speed and ability to cover vast areas that are difficult to access by ground crews.

6.2.5.1 Aerial suppression

Aerial suppression employs diverse aircraft types and methods, each tailored to specific wildfire scenarios. Fixed-wing aircraft, such as the DC-10 or the Bae 146, are employed for the dissemination of substantial quantities of water or fire retardants over expansive areas. Their effectiveness lies in the creation of firebreaks and the protection of extensive regions of land or infrastructure.

Helicopters, such as the Sikorsky S-64 or the Bell 205, possess versatility and are adept at delivering water or retardants with great precision. They are frequently utilized for targeting specific hot spots or smaller areas where fixed-wing aircraft may encounter limitations.

Aerial suppression offers numerous advantages. It enables rapid coverage of vast and remote areas that are challenging to access by ground, thereby facilitating the timely management of widespread fires and preventing their further dissemination.

Speed is another key advantage. Aircraft can swiftly deploy water or retardants, providing immediate assistance in controlling the fire and safeguarding vulnerable areas. This speed is paramount in managing rapidly evolving wildfires.

Furthermore, aerial suppression enhances accessibility to areas that are otherwise inaccessible due to rugged terrain, dense vegetation, or other obstacles. This accessibility facilitates the management of fires in challenging environments.

However, there are also certain limitations. The effectiveness of aerial suppression is significantly compromised by adverse weather conditions, such as high winds, poor visibility, or extreme heat. These factors can impair the accuracy and safety of aerial operations.

The utilization of aircraft in firefighting necessitates meticulous coordination with ground crews and other aerial operations to minimize the risk of accidents and ensure effective deployment. The safety of pilots and the potential for mid-air collisions must be diligently managed.

Concurrently, operating and maintaining firefighting aircraft incurs substantial costs. These costs encompass fuel, maintenance, and personnel, which can be substantial compared to alternative suppression methods.

6.2.5.2 Water bombing

Water bombing is a specific form of aerial suppression that involves dropping large quantities of water onto the fire to cool it and extinguish flames (Graham, 2020; Underwood and Held, 2012). Various methods and equipment are employed for water bombing operations. Large, fixed-wing aircraft equipped with water tanks can release substantial quantities of water in a single pass. These tankers are specifically designed to cover vast areas swiftly and efficiently.

Helicopters can carry and drop water using large buckets, often referred to as "bambi buckets" (Kal'avský et al., 2019; Matta, 2020). This method is particularly useful for precise application and targeting specific hotspots.

Water bombing offers several advantages. Directly cooling the fire by reducing the temperature of the burning materials is crucial for extinguishing flames and controlling the fire's intensity.

Water bombing also provides an immediate impact on the fire, effectively controlling its spread and protecting critical areas. This rapid response is vital for managing fast-moving fires. Additionally, the method exhibits flexibility. Helicopters equipped with buckets can perform precise water drops, enabling targeted suppression of hot spots and smaller fire areas. This flexibility enhances the overall effectiveness of wildfire management.

However, water bombing also faces certain limitations. The effectiveness of water bombing is contingent upon the availability of nearby water sources. In regions with limited water supply, the frequency and volume of drops may be constrained.

While water bombing provides a short-term solution by cooling the fire, it does not necessarily address the underlying issues of fuel accumulation or fire behavior. Consequently, it may require the integration of other suppression methods for long-term control.

Furthermore, there is a significant potential for runoff, which may lead to soil erosion and adversely impact water quality in nearby rivers and streams. Therefore, the environmental impact of large-scale water bombing necessitates careful consideration before deployment.

6.2.5.3 Integration and coordination

Aerial suppression and water bombing are most effective when integrated with other wildfire suppression methods. Coordination between aerial operations and ground crews is essential to ensure that resources are used efficiently and that different suppression strategies complement each other. For example, aerial suppression can be used to control the spread of the fire, while ground crews work to extinguish hot spots and create firebreaks.

6.2.6 FIRE SUPPRESSION TEAMS

Fire suppression teams are indispensable in managing and extinguishing wildfires. Comprising a diverse array of specialized personnel, these teams collaborate effectively to control and contain fires, minimize damage, and prioritize safety. Their effectiveness relies on coordination, training, and the strategic use of resources to address the complex and dynamic nature of wildfires (McLennan et al., 2006).

6.2.6.1 Composition

Wildland firefighters are the frontline personnel in wildfire suppression (Desmond, 2006). Wildland firefighters are trained to address the specific challenges posed by wildfires, such as remote locations and extreme environmental conditions. They are organized into three distinct crews: hotshot crews, helitack crews, and smokejumpers.

- **Hotshot crews:** These are elite, highly trained teams of wildland firefighters who tackle the most challenging fires (Bramwell, 2008). They possess expertise in constructing firebreaks, executing controlled burns, and implementing intricate suppression strategies.
- **Helitack crews:** Helitack crews are specialized in using helicopters for fire suppression (Page and Butler, 2019). Their primary functions encompass aerial reconnaissance, water or retardant drops, and the transportation of personnel and equipment.

- **Smokejumpers:** Smokejumpers are firefighters who are trained to parachute into remote areas to quickly address wildfires (Beil, 1999; Taylor, 2014). They possess expertise in rapid deployment and initial fire suppression operations in challenging and inaccessible environments.

Support personnel provide critical assistance to the firefighting teams, ensuring that operations run smoothly and efficiently. They are categorized into fire behavior analysts, logistics and supply specialists, and medical personnel.

- **Fire behavior analysts:** These experts assess the behavior of the fire, including its spread, intensity, and potential impact (Adkins et al., 2021; Molina et al., 2007). Their analysis facilitates the formulation of well-informed decisions regarding fire suppression tactics and strategies.

Logistics and supply specialists: These professionals oversee the logistical aspects of a firefighting operation, including the procurement and distribution of equipment, supplies, and sustenance. They ensure that fire suppression teams have the requisite resources to execute their tasks effectively.

- **Medical personnel:** Medical teams, including paramedics and doctors, provide emergency medical care to injured firefighters and support overall health and safety on the ground.

The **Incident Command System (ICS)** is a standardized approach to managing emergency incidents, including wildfires (Jensen and Thompson, 2016; Wang et al., 2012). This position encompasses a diverse range of responsibilities, including the roles of incident commander, operations section chief, planning section chief, logistics section chief, and finance/administration section chief.

- **Incident commander:** The incident commander is responsible for overall fire management and decision-making (Rake and Njå, 2009). They collaborate with various teams and agencies to ensure a coordinated and effective response.

- **Operations section chief:** They oversee the tactical operations of the firefighting teams, including the deployment of resources and execution of suppression strategies (Deal et al., 2006; Hunter, 1993).

- **Planning section chief:** The planning section chief is responsible for developing and maintaining the incident action plan, including objectives, strategies, and resource allocation (Cadd et al., 1999; Deal et al., 2006).

- **Logistics section chief:** This individual manages the logistical needs of the firefighting operation, including transportation, supplies, and equipment.

- **Finance / administration section chief:** They handle financial and administrative aspects, such as budgeting, cost tracking, and personnel management.

6.2.6.2 Training and preparedness

Fire suppression teams undergo rigorous training to prepare for the demands of wildfire response (Dorrer et al., 2018; Tedim et al., 2020b). Training programs include firefighter training, safety training, and specialized training.

- **Firefighter training:** Basic and advanced training programs cover fire behavior, safety protocols, suppression techniques, and use of equipment. Firefighters acquire the expertise to operate effectively in diverse environments and respond appropriately to emergency situations.
- **Safety training:** Ensuring the safety of firefighting personnel is a priority. Training encompasses a comprehensive curriculum that includes first aid, survival techniques, and safety protocols. These components are designed to mitigate potential risks and safeguard the well-being of team members.
- **Specialized training:** Teams like hotshot crews, helitack crews, and smokejumpers receive specialized training for their specific roles. This encompasses advanced tactical maneuvers, aerial operations, and parachuting proficiency.

6.2.6.3 Coordination and communication

Effective coordination and communication are paramount for the success of fire suppression operations. Wildfire suppression frequently necessitates collaboration among diverse agencies, encompassing local, state, and federal organizations. Coordinated efforts optimize resource allocation and ensure comprehensive coverage of all response components.

Reliable communication systems are also crucial for maintaining contact between various teams and the incident command. This includes radios, satellite phones, and other communication tools that facilitate real-time information sharing.

Additionally, regular briefings and meetings are conducted to provide teams with updates on fire conditions, strategies, and safety information. These briefings ensure that all stakeholders are adequately informed and aligned with the incident action plan.

6.2.7 MONITORING OF WILDFIRE SUPPRESSION

Monitoring is a critical aspect of wildfire suppression that ensures effective management and control of wildfires. It involves tracking the progress, behavior, and effectiveness of firefighting efforts to make informed decisions, adapt strategies, and efficiently utilize resources. Monitoring provides real-time data that aids in assessing fire conditions, evaluating suppression tactics, and ensuring the safety of personnel.

6.2.7.1 Need for monitoring

Effective monitoring is paramount for several reasons. It enables the early detection of wildfires and their progression. By identifying the fire's location, size, and intensity promptly, firefighting teams can respond swiftly, mitigating the risk of the fire escalating uncontrollably. Early detection is also crucial for implementing timely suppression measures and preventing extensive damage.

Monitoring also facilitates the assessment of fire behavior, which is essential for effective suppression. Monitoring provides data on the fire's spread, intensity, and the impact of weather conditions. This information enables fire behavior analysts to forecast future fire behavior and formulate appropriate suppression tactics.

Continuous monitoring also aids in evaluating the efficacy of suppression efforts. By tracking changes in fire size and intensity, teams can ascertain whether current strategies are effective or if adjustments are necessary. This feedback loop ensures that resources are allocated judiciously and that suppression strategies are adapted as required.

Furthermore, monitoring data guides the allocation of resources by identifying areas that necessitate increased attention or additional support. For instance, if a fire is spreading rapidly in a specific region, additional firefighting resources can be directed to that location to address the threat. This targeted approach ensures that resources are deployed where they are most impactful.

Monitoring also plays a pivotal role in safeguarding the safety of firefighting personnel. By tracking fire behavior and identifying potential hazards, incident commanders can make informed decisions that protect crews and minimize risks. Real-time data facilitates the assessment of the safety of various areas and the planning of safe evacuation routes, if necessary.

Strategic suppression requires meticulous planning based on reliable data. Monitoring provides insights into fire conditions and facilitates the development and refinement of suppression strategies. This encompasses the creation of firebreaks, conducting controlled burns, and managing fuel loads. Strategic planning informed by monitoring enhances the overall effectiveness of wildfire management.

6.2.7.2 Methods of monitoring

There are several methods of monitoring. Often a combination of them is used simultaneously. **Ground-based monitoring** involves direct observation and data collection from the fire scene. **Fire lookouts** are positioned in elevated locations to observe large areas for signs of fire. They provide visual confirmation of fire activity and report observations to incident command centers. **Firefighters and field crews** also continuously monitor fire behavior and assess the effectiveness of suppression techniques. They check for new hotspots, flare-ups, and changes in fire intensity, providing valuable feedback for tactical adjustments. Instruments such as **thermal**

imaging devices are often used by ground crews to detect heat sources and monitor temperature changes, helping identify hot spots that require additional attention.

Remote monitoring technologies offer advanced solutions for tracking wildfire activity. Optical and infrared **fire detection cameras** can detect smoke and fire from a distance, providing real-time alerts and visual confirmation of fire activity. Similarly, **automated detection systems** combine data from various sensors and algorithms to detect anomalies such as smoke or heat, offering early warnings and continuous monitoring. **Weather stations** monitor environmental conditions such as temperature, humidity, wind speed, and precipitation, providing data crucial for understanding fire behavior.

Aerial monitoring can provide a broader bird's-eye perspective and help in rapid data collection. **Aircraft surveillance** with fixed-wing aircraft and helicopters equipped with cameras and sensors offers real-time imagery and data on fire size, intensity, and spread. This aerial view helps in assessing the overall fire situation and guiding suppression efforts. **Drones** capture high-resolution images and thermal data, offering detailed insights into fire behavior and hot spots. They are useful for inspecting areas that are difficult to access.

Satellites with thermal imaging sensors similarly provide comprehensive coverage of fire activity, tracking progression, burn severity, and perimeter changes. Satellite data facilitates the delineation of fire perimeters and the evaluation of their impact.

These inputs are used by **fire behavior analysts** to predict fire behavior and inform suppression strategies by analyzing factors such as fire spread rates, intensity, and weather patterns.

6.2.7.3 Integration with suppression efforts

Effective monitoring is a fundamental component of optimizing wildfire suppression efforts.

- **Incident Command Coordination:** Data monitoring enables incident commanders to receive timely information regarding fire conditions, resource requirements, and tactical adjustments. Real-time data facilitates effective coordination among ground crews, aerial teams, and supporting personnel.
- **Resource Management:** Monitoring guides the allocation of firefighting resources, ensuring they are deployed where they are most needed based on current fire conditions.
- **Strategy Adjustment:** Continuous monitoring enables real-time adjustments to suppression strategies, enhancing their effectiveness and accommodating evolving fire behavior.
- **Safety:** Monitoring ensures the safety of personnel by tracking fire behavior and identifying hazards, enabling safe operations and evacuation plan-

ning.

6.3 DATA AND WILDFIRE SUPPRESSION

In the context of wildfire suppression, data serves as a pivotal asset that supports various facets of fire management. Efficient utilization of data enables firefighting teams to acquire a comprehensive understanding of fire behavior, optimize resource allocation, and adapt suppression strategies in a timely manner. Systematic collection, analysis, and application of data are indispensable for the development of effective wildfire suppression tactics and the enhancement of overall fire management outcomes.

6.3.1 TYPES OF DATA USED IN WILDFIRE SUPPRESSION

Wildfire suppression requires several kinds of data. **Fire behavior data** provides insights into the dynamic nature of wildfires, including their spread, intensity, and interaction with environmental factors (Albini, 1976a; Artés et al., 2019; Silva et al., 2022). This type of data is critical for predicting fire behavior and planning appropriate suppression measures. This includes:

- **Fire spread rates:** The rate at which a fire spreads is a key metric for assessing the urgency and scale of the response required. This data is collected through observations and satellite imagery and helps determine how quickly a fire is advancing (Monedero et al., 2019; Pereira et al., 2022; Perrakis et al., 2014). Understanding the spread rates of wildfires is crucial for planning resource allocation and forecasting future fire behavior.
- **Intensity and heat levels:** Data on the intensity of the fire and the heat it generates informs decisions about suppression tactics (Liu et al., 2010). High-intensity fires necessitate more aggressive suppression strategies, whereas lower-intensity fires may be managed utilizing alternative approaches. Additionally, heat level data is employed to evaluate the potential for spot fires and embers, which can further disseminate the fire.
- **Weather conditions:** Weather plays a significant role in influencing fire behavior. Temperature, humidity, wind speed, and wind direction significantly influence the propagation and escalation of wildfires. Through meticulous analysis of weather conditions, firefighting teams can effectively anticipate alterations in fire behavior and modify their suppression strategies accordingly.

Satellite and aerial data provide a broad and detailed perspective on wildfire activity, which is essential for large-scale monitoring and strategic planning.

- In **satellite imagery,** satellites equipped with thermal and optical sensors offer comprehensive views of wildfire activity. Thermal imaging detects heat emitted by the fire, while optical sensors provide visual images. This data is utilized to delineate fire perimeters, assess burn severity, and monitor the temporal progression of the fire. Satellite imagery also facilitates the identification of active fire fronts and the assessment of regions that may require supplementary resources.

- In **aerial surveys,** aircraft, including fixed-wing planes and helicopters, are equipped with cameras and sensors to capture real-time images and data. Aerial surveys provide comprehensive visualizations of the fire's current state, facilitate the evaluation of the efficacy of suppression strategies, and direct the allocation of firefighting resources. Drones, in particular, offer high-resolution imagery and thermal data that are highly valuable for inspecting inaccessible regions and providing precise information on fire behavior.

Ground-based observations involve collecting data directly from the fire scene, providing immediate and detailed insights into fire conditions. These include:

- **Fire lookouts:** Fire lookouts are strategically positioned in elevated locations to observe large areas and report on fire activity (Berryoung, 2021). These real-time visual observations of fire size, intensity, and behavior provide valuable insights into the fire's status. Their reports contribute to a comprehensive understanding of the fire's evolution and facilitate the coordination of response efforts.
- **Firefighters and field crews:** Ground crews collect data on fire behavior, suppression effectiveness, and environmental conditions. Their observations encompass monitoring fire intensity, identifying novel hotspots, and evaluating the impact of suppression strategies. This data is pivotal for making on-the-ground adjustments and enhancing suppression efforts.
- **Thermal imaging devices:** Ground crews use thermal imaging devices to detect heat sources and monitor temperature changes. These devices assist in identifying regions with elevated temperatures, suggesting potential hotspots or areas necessitating enhanced suppression measures. Thermal imaging proves particularly valuable in detecting smoldering areas that may not be discernible but could potentially reignite.

6.3.2 INTEGRATION OF DATA INTO SUPPRESSION STRATEGIES

Integrating data into wildfire suppression strategies significantly enhances the effectiveness of fire management efforts by providing actionable insights and supporting informed decision-making. This entails:

- **Strategic planning:** Data serves as a crucial input in the development of suppression strategies. This includes the creation of firebreaks, controlled burns, and the management of fuel loads. Accurate data on fire behavior and environmental conditions enables the design of interventions that effectively address current fire conditions and anticipate potential changes.

- **Resource allocation:** Data-driven resource allocation enables effective firefighting operations by prioritizing areas with the highest risk. For instance, when data reveals rapid fire spread in a particular region, additional resources can be swiftly deployed to mitigate the threat. This targeted approach ensures that resources are deployed where they are most impactful, optimizing their overall effectiveness.
- **Real-time adjustments:** Continuous data collection enables real-time adjustments to suppression strategies. As fire behavior evolves, data facilitates modifications to tactics and resource deployment. This adaptability is crucial for responding to dynamic fire conditions and enhancing suppression effectiveness.
- **Risk assessment:** Data is instrumental in assessing risks and making decisions that safeguard lives, property, and the environment. Through the analysis of fire behavior and environmental conditions, incident commanders can identify potential hazards, implement mitigation measures, and devise safe evacuation routes, should they become necessary.

6.4 FACTORS DETERMINING THE CHOICE OF METHODS

Selecting the appropriate wildfire suppression methods is paramount for effective fire management. The selection process is influenced by a multitude of factors that impact the efficiency, safety, and success of the suppression effort. These factors encompass fire behavior, environmental conditions, available resources, suppression objectives, and safety considerations. By comprehending these variables, a strategic approach to wildfire suppression can be developed that maximizes effectiveness and minimizes risk.

6.4.1 FIRE BEHAVIOR

Fire behavior includes a consideration of several factors, including:

- **Fire intensity and size:** The intensity and size of the wildfire are primary factors in determining suppression methods. High-intensity fires, characterized by intense heat and rapid spread, necessitate robust suppression techniques. These techniques include aerial water drops, large-scale firebreaks, and extensive ground operations. For instance, wildfires burning at high intensity may necessitate the use of heavy machinery to construct firebreaks and large aerial tankers for water or retardant drops. Conversely, smaller fires may be managed with less aggressive methods, such as using hand crews and localized water application.
- **Fire spread rate:** The rate at which a fire spreads affects the choice and immediacy of the suppression response. Rapidly spreading fires pose a heightened risk of rapid escalation and necessitate prompt and extensive

suppression efforts. This may entail the rapid deployment of additional resources and the implementation of aggressive tactics to impede the fire's progression. Conversely, slower-moving fires may be managed with more deliberate tactics, enabling a more measured approach to suppression and resource allocation.

- **Fire behavior patterns:** Analyzing fire behavior patterns, such as the direction of spread, intensity fluctuations, and the occurrence of spot fires, is essential for selecting suppression methods. For instance, if a fire is propelled by strong winds and moving in a predetermined direction, suppression efforts may concentrate on establishing firebreaks and deploying resources along the anticipated path of the fire. By comprehending these patterns, it becomes possible to anticipate alterations in fire behavior and adjust suppression strategies accordingly.

6.4.2 ENVIRONMENTAL CONDITIONS

Environmental conditions encompass a comprehensive analysis of terrain, topography, weather patterns, and the availability and characteristics of fuel resources. The physical characteristics of the **terrain** significantly influence the choice of suppression methods. Terrain that is steep, rugged, or inaccessible may restrict the application of certain suppression techniques. In such instances, ground crews may require specialized equipment, such as chainsaws and hand tools, to establish firebreaks. Conversely, terrain that is flat or easily accessible permits the utilization of larger-scale equipment, such as bulldozers, and facilitates more extensive ground operations.

Weather conditions play a critical role in fire behavior and the effectiveness of suppression efforts. Elevated temperatures and reduced humidity can significantly enhance fire behavior, necessitating the deployment of more aggressive suppression strategies. Wind speed and direction also play a crucial role in fire propagation and can influence the effectiveness of various tactics. For instance, strong winds can rapidly spread the fire, bypassing containment lines. Consequently, aerial support and additional ground resources are often required to mitigate the challenges posed by wind-driven fire behavior.

The **type and amount of available fuel** also significantly impact the choice of suppression methods. Dense vegetation or heavy fuel loads intensify and spread wildfires, necessitating more intensive suppression tactics such as controlled burns to reduce fuel loads or extensive water drops to cool the fire. Conversely, areas with lighter fuel loads can be managed with less aggressive approaches. Understanding fuel characteristics aids in selecting suitable methods to effectively control and contain wildfires.

6.4.3 RESOURCE AVAILABILITY

The availability of **firefighting personnel and equipment** is a crucial factor in determining the choice of suppression methods. Large-scale fires necessitate the mobilization of additional personnel and specialized equipment, including bulldozers, water tenders, and aerial tankers. The effective deployment and coordination of these resources significantly influence the selection of suppression tactics. For instance, in cases requiring a comprehensive response, incident commanders may opt for a combination of aerial and ground-based methods to effectively address the fire's demands.

Financial constraints and logistical considerations also impact the selection of suppression methods. Methods that necessitate substantial resources, such as extensive aerial drops or substantial ground crew deployments, may be constrained by budgetary limitations. Logistical considerations, including the accessibility of transportation routes and the necessity for logistics, also influence the feasibility of certain suppression tactics. Efficient planning and resource management are paramount to ensure that suppression efforts are cost-effective and practical within the confines of available resources.

6.4.4 OBJECTIVES OF SUPPRESSION

Firefighting operations can be conducted with various objectives. The selection of fire suppression method will be contingent upon the specific objective of suppression. When aiming to **protect lives and property,** suppression methods are chosen based on their effectiveness in safeguarding residential areas, critical infrastructure, and natural resources. Strategies for managing wildfires may include constructing firebreaks around communities, evacuating residents, and prioritizing resource deployment to high-risk areas. The objective is to minimize the impact on human life and property while effectively controlling the fire.

When targeting to **minimize environmental impact,** suppression efforts must balance fire control with minimizing environmental impact. Controlled burns, for instance, can effectively manage fuel loads and reduce fire intensity but need to be meticulously planned to mitigate adverse effects on ecosystems. The selection of suppression methods is guided by the potential environmental consequences, with the objective of achieving effective fire management while preserving natural habitats and minimizing ecological disruption.

Another objective is to **contain the fire** within predetermined boundaries. In this regard, suppression methods are chosen based on their capacity to prevent the fire from spreading beyond containment lines. Techniques such as constructing firebreaks, applying fire retardants, and utilizing aerial suppression are employed to achieve containment. Effective containment strategies entail coordinating resources and tactics to maintain control over the fire's perimeter and prevent further expansion.

6.4.5 SAFETY CONSIDERATIONS

Ensuring the **safety of firefighting personnel** is a critical consideration in choosing suppression methods. The selected methods must ensure the safety of crews from hazardous conditions and minimize the risk of injury. For instance, aerial suppression methods may be employed to reduce direct exposure to the fire, while ground crews are provided with appropriate protective gear and safety protocols. The safety of personnel is of utmost importance, and suppression strategies are continually adjusted to address potential risks and hazards.

Public safety is equally a crucial consideration in suppression efforts. This includes ensuring that suppression methods do not endanger nearby communities or first responders. Techniques that involve controlled burns or large-scale water drops must be managed meticulously to prevent unintended consequences such as the spread of embers or flooding. Clear communication with the public and effective evacuation plans are essential components of maintaining safety during wildfire suppression.

7 Post-fire management

Post-wildfire management is paramount in addressing the extensive and multifaceted impacts of wildfires. It encompasses a range of strategies aimed at mitigating damage, restoring ecosystems, supporting communities, and preparing for future events. By implementing effective post-wildfire management, organizations can effectively minimize long-term damage and expedite the recovery process. This chapter delves into the critical components of post-wildfire management, including damage assessment, hazard mitigation, ecosystem restoration, and community support. It also integrates key facts about post-fire restoration and the Burned Area Emergency Response (BAER) to provide a comprehensive view of the challenges and ongoing efforts. Furthermore, it underscores that post-wildfire restoration entails both time-consuming and costly endeavors, necessitating substantial resources and a long-term commitment.

7.1 LID AND BIOLOGICAL COMMUNITIES

Large infrequent disturbances (LID), such as large wildfires, floods, and volcanic eruptions, play a significant role in shaping biological communities and ecosystems. Although less frequent, these disturbances have significant impacts on ecological structures, processes, and species interactions. Understanding how these disturbances influence biological communities is crucial for managing ecosystems, conserving biodiversity, and predicting future ecological changes.

7.1.1 CHARACTERISTICS OF LID

Large infrequent disturbances are characterized by their rarity and substantial impact on the environment. In contrast to frequent disturbances, which may occur regularly, large infrequent disturbances occur at longer intervals but with greater intensity. Examples include major wildfires, severe floods, and large-scale volcanic eruptions.

The magnitude of these disturbances is typically substantial, affecting large areas and causing widespread changes to the landscape. Their high intensity can lead to severe alterations in habitat structure, soil composition, and water systems. For instance, a large wildfire can completely consume vegetation over vast areas, significantly altering soil properties and hydrological patterns.

Due to their scale and severity, large infrequent disturbances can cause significant ecological shifts. They often result in immediate and dramatic changes to the environment, followed by a complex process of ecological recovery and succession. These disturbances can lead to the loss of species, alteration of habitat conditions, and changes in ecosystem functions.

7.1.2 EFFECTS ON BIOLOGICAL COMMUNITIES

Significant and infrequent disturbances can have substantial impacts on biological communities. They can drastically **alter species composition and diversity** within biological communities. Some species may be eliminated or reduced in number, while others may benefit from the altered conditions. For instance, certain fire-adapted plant species may flourish after a wildfire, while others may be severely impacted or displaced.

The **alteration of habitats** due to large disturbances can create new ecological niches and opportunities for species. Disturbed areas may undergo a process of ecological succession, wherein new plant and animal communities gradually replace those lost during the disturbance. This process can lead to alterations in habitat types and modifications in the distribution of species.

Ecosystem processes such as nutrient cycling, water regulation, and energy flow can also significantly be impacted by large infrequent disturbances. For instance, a significant flood can disrupt nutrient distributions and alter water quality, which, in turn, influences plant and animal communities.

7.1.3 ECOLOGICAL SUCCESSION AND RECOVERY

Following a large disturbance, ecosystems typically undergo processes of ecological succession. Ecological succession is defined as "the process of change in the species structure of an ecological community over time (Awadhiya, 2021)." In this process, an ecosystem changes, often from one that supports minimal (or no) life forms (also known as pioneer community), to one that is stable and supports a large biodiversity (also known as climax community). In areas where the disturbance has removed all existing soil and vegetation, we get primary succession, leading to a gradual establishment of new plant communities. In areas where some soil and vegetation remain, secondary succession occurs, allowing for a faster recovery of plant and animal communities.

In the early stages of succession, pioneer species play a crucial role. These species are generally resilient and adept at colonizing disturbed areas swiftly. They contribute to soil stabilization, enhance soil conditions, and foster a more favorable environment for other species to establish.

The recovery process from infrequent, large-scale disturbances can be protracted and encompasses several stages of ecological succession. Over time, ecosystems may revert to a state akin to their pre-disturbance conditions, albeit this process can span decades or even centuries. The rate and trajectory of recovery are contingent upon the severity of the disturbance, the ecosystem's resilience, and ongoing environmental factors. In certain instances, it may be possible for the original pre-disturbance conditions to remain unattainable even after several centuries have elapsed.

7.1.4 IMPLICATIONS FOR ECOSYSTEM MANAGEMENT

This comprehension holds significant implications for ecosystem management practices. Understanding the role of large infrequent disturbances in shaping biological communities is important for **developing effective conservation strategies**. Management practices should consider the natural disturbance regime of an ecosystem and aim to maintain or restore ecological processes.

Conservation efforts should also focus on **enhancing the adaptability of species and ecosystems** to changing conditions. This may entail safeguarding vital habitats, restoring degraded regions, and fostering biodiversity to enhance ecosystem resilience.

In this context, **predictive modeling** can be used to anticipate the potential impacts of large disturbances on biological communities. Through the simulation of diverse disturbance scenarios, researchers and managers can gain a comprehensive understanding of the potential outcomes of disparate interventions. This knowledge enables them to formulate strategies to mitigate adverse effects and facilitate a swift recovery.

7.2 POST-FIRE MANAGEMENT OPTIONS

Several post-wildfire operations are available, with habitat improvement strategies tailored to the specific condition and requirements of the affected area. Some of the strategies are detailed below [Figure 7.1] (Awadhiya, 2021):

1. **Recovery / Neglect:** This approach involves allowing natural processes to take their course without active intervention. Over time, the disturbed habitat will undergo ecological succession as native species gradually recolonize the area. For instance, leaving a burned forest to recover naturally may eventually result in the area being re-vegetated by pioneer species. This method is cost-effective and relies on natural recovery processes. However, it can be a slow process, and there is a risk that the habitat could degrade further if left neglected, such as when erosion exacerbates soil loss. Additionally, the area may be colonized by invasive species, leading to the loss of habitat function for native species.

2. **Restoration:** This strategy focuses on actively returning the habitat to its original state, one that existed prior to the disturbance. In the aftermath of a severe wildfire that has devastated a forest, restoration efforts will encompass a range of activities, including soil and moisture conservation, and the replanting of native species that were once present in the ecosystem. The primary objective is to restore the pre-disturbance conditions as

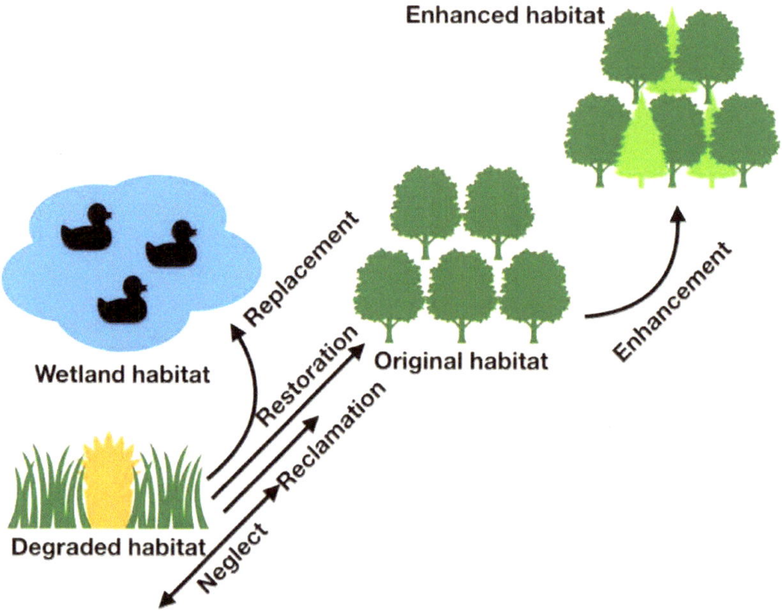

FIGURE 7.1: Some strategies for habitat improvement.

closely as feasible, thereby restoring ecological balance and facilitating the survival of native wildlife.

3. **Rehabilitation / Reclamation:** This approach seeks to improve the degraded habitat to a condition of higher ecological value, which may not necessarily be identical to the original state. For instance, after a wildfire, instead of attempting to restore the forest exactly as it was, rehabilitation might involve reforestation with species adapted to the new conditions or even fast-growing exotic species. This method is often chosen when restoration is too challenging or resource-intensive. Rehabilitation can serve as an interim solution, with the potential for more comprehensive restoration later when resources permit.

4. **Enhancement:** This strategy aims to increase the ecological value of the habitat by adding features that benefit wildlife. For instance, constructing water holes or wildlife shelters in a post-wildfire landscape can enhance conditions for animals and support their recovery. Enhancement is typically employed when the habitat is not severely degraded, and this strategy can be employed as a targeted way to improve specific aspects of the environment.

5. **Replacement:** This approach involves creating a new type of habitat in place of the degraded one. For instance, a heavily eroded forest area may be replaced with a marshy wetland. Replacing the original habitat can be more practical and cost-effective when the original habitat is too damaged or difficult to restore. However, it is crucial to ensure that the new habitat does not introduce new risks to wildlife. For example, necessary treatment to avoid toxic leaching is essential before transforming an area into a wetland.

Each of these strategies presents its own set of advantages and limitations. The selection of approach is contingent upon the magnitude of the damage, the availability of resources, and the long-term management objectives. Effective post-wildfire habitat management frequently entails a combination of these strategies to address both immediate and long-term ecological requirements.

7.3 POST-WILDFIRE MANAGEMENT STEPS

Post-wildfire management is a multi-step process that involves assessing the extent of damage caused by the wildfire. This assessment typically utilizes techniques such as normalized burn ratio, ground surveys, and aerial surveys. Subsequently, immediate hazards such as soil erosion are mitigated, and emergency responses are provided. Following these measures, efforts are made to restore the ecosystem and community infrastructure, often with the active participation of local communities. Finally, long-term recovery plans are formulated and implemented.

7.3.1 DAMAGE ASSESSMENT

A thorough assessment of wildfire damage encompasses a comprehensive evaluation of both environmental and socioeconomic impacts. Initially, aerial surveys and satellite imagery aid in identifying burned areas, while ground assessments provide insights into the destruction of vegetation, wildlife habitats, and soil integrity. Furthermore, the evaluation extends to human infrastructure, including homes, businesses, and essential services, which can be severely affected. Economic losses are quantified through property damage estimates and disruptions to local economies, particularly in tourism and agriculture.

7.3.1.1 Assessment of burnt areas

The initial step in post-wildfire management is to assess the extent and severity of the damage. This involves evaluating the burned areas, the intensity of the fire, and its impact on vegetation, soil, and infrastructure. Teams utilize aerial surveys, satellite imagery, and ground inspections to gather comprehensive data. This assessment provides pivotal information for planning recovery and restoration endeavors. Notably, wildfire damage can persist beyond the immediate aftermath of the flames. Loss of vegetation can expose soil to erosion, potentially leading to severe consequences such as floods and debris flows that may damage structures, roads, trails, and water reservoirs, and jeopardize community water supplies.

The Normalized Burn Ratio (NBR) is a critical index utilized in the context of burnt area assessment, offering valuable insights into both the current state of vegetation and the effectiveness of various measures. NBR is calculated using data from two specific satellite spectral bands: near-infrared (NIR) and shortwave infrared (SWIR). The formula for NBR is:

$$\text{NBR} = \frac{(\text{NIR} - \text{SWIR})}{(\text{NIR} + \text{SWIR})}$$

This index utilizes the high reflectance of healthy vegetation in the near-infrared (NIR) band and the high reflectance of burnt areas in the short-wave infrared (SWIR) band. By comparing these bands, the NBR provides a clear visualization of vegetation conditions and changes resulting from wildfires, making the index indispensable for wildfire management strategies.

Following a wildfire, NBR can be employed to monitor post-fire recovery and assess fire damage. By comparing post-fire NBR values with pre-fire data, stakeholders can assess the extent of vegetation loss and evaluate the efficacy of recovery efforts. This information is also pivotal in comprehending the regeneration rate of an area and determining whether additional measures are necessary to mitigate future fire risks.

The ratio is also employed for the identification of high-risk areas for future wildfires. By analyzing historical NBR data, in conjunction with current vegetation and moisture conditions, fire management agencies can pinpoint regions that are particularly susceptible to wildfires. Areas with lower NBR values, indicating reduced vegetation health or increased dryness, are often more prone to ignition and rapid fire spread. This information facilitates targeted preventive measures, such as vegetation thinning or controlled burns, to reduce fuel loads and diminish fire risk in vulnerable areas.

NBR also provides valuable insights into fuel loads and vegetation health, which are crucial factors in future wildfire prevention. High fuel loads and unhealthy vegetation contribute to the intensity and spread of wildfires. Regular monitoring of NBR values enables fire management professionals to assess changes in vegetation health and fuel accumulation over time. This ongoing evaluation facilitates the planning and implementation of proactive measures, such as removing excess vegetation or adjusting land management practices to mitigate fire risk.

NBR is also a valuable tool for evaluating the efficacy of wildfire prevention strategies. Following the implementation of measures such as controlled burns or vegetation management, satellite imagery can be utilized to assess alterations in NBR values. An elevation in NBR, indicative of enhanced vegetation health or diminished dry fuel loads, suggests that the preventive measures are exhibiting a positive impact. Conversely, stable or diminishing NBR values may necessitate adjustments or supplementary interventions to further mitigate fire risk.

The insights provided by NBR can also support community preparedness and planning. By identifying high-risk areas and assessing vegetation health, communities can develop targeted fire prevention plans and educational campaigns. For instance, data-driven recommendations can guide homeowners in creating defensible space around properties and adopting fire-safe practices.

7.3.1.2 Infrastructure and property damage

Evaluating the condition of infrastructure and property is crucial for prioritizing repair and reconstruction efforts. This assessment encompasses the assessment of roadways, bridges, utilities, and buildings. Accurate assessments facilitate the identification of immediate needs and ensure the restoration of essential services to support impacted communities. By prioritizing efforts and resources, this evaluation helps address the most pressing needs.

This assessment is typically conducted through ground surveys, aerial surveys, and drone-based surveys.

7.3.1.3 Environmental impact

Evaluating the environmental impact involves understanding changes to soil quality, water resources, and wildlife habitats. Wildfires can cause soil erosion, water contamination, and habitat loss. Identifying these impacts is essential for developing strategies to mitigate environmental damage and support ecological recovery. For instance, erosion and sedimentation can persist in ecosystems for years after a wildfire has been extinguished.

Environmental impact evaluations are typically conducted as research studies. The knowledge gained from assessments of past wildfires is often applied to the current wildfire. In some priority areas, fresh *ab initio* environmental impact studies may also be conducted.

7.3.2 HAZARD MITIGATION

Following a wildfire, several novel hazards emerge, including extensive erosion, flooding risks in downstream regions, and the further dissemination of hazardous substances generated by the wildfire, including into the groundwater. Consequently, the mitigation of these hazards necessitates a prioritized approach.

- **Erosion control:** Areas with lost vegetation are particularly vulnerable to erosion, which can lead to further damage such as landslides, siltation, and loss of soil. Erosion control measures, including the planting of ground cover, the installation of silt fences, and the construction of check dams, are employed to stabilize the soil and prevent sedimentation in water bodies. These measures serve to protect the landscape and mitigate the risk of floods and landslides. Soil stabilization techniques, such as reseeding with native grasses, replanting trees and shrubs, and employing mechanical methods, are essential in preventing ongoing erosion and damage.

- **Flood risk management:** The risk of flooding increases after a wildfire due to the loss of vegetation and altered soil properties. Flood risk management strategies encompass the construction of retention basins, the reinforcement of stream banks, and the enhancement of drainage systems. Effective flood management serves as a crucial measure in mitigating the potential damage caused by post-wildfire floods and ensuring the safety of communities. This is particularly important given the significant impact that the loss of vegetation can have on water flow patterns and the subsequent increase in the likelihood of flood events.

- **Hazardous materials removal:** Wildfires can leave hazardous materials in the environment, such as debris from burned structures and chemical contaminants. The removal of these materials is paramount for ensuring public safety and environmental health. Specialized teams are responsible for the cleanup and safe disposal of hazardous waste to prevent further harm. Addressing hazardous materials is a crucial component of protecting both human health and the environment in the aftermath of a wildfire.

7.3.3 BURNED AREA EMERGENCY RESPONSE (BAER)

As soon as it is safe, the Forest Service initiates a Burned Area Emergency Response (BAER) to address immediate post-fire conditions (Tracy and Ruby, 1994; Witt, 1999). BAER assessments aim to protect life, property, water quality, and ecosystems from further damage (Olsen, 2007). This proactive approach is paramount in mitigating the ongoing effects of the wildfire and preventing further harm.

The primary objectives of BAER are to stabilize soil, control water and sediment movement, and protect vital resources. In the event of emergency conditions following a fire, BAER actions may entail immediate measures such as installing erosion control measures, stabilizing slopes, and managing debris flows. The objective is to alleviate emergency conditions and prevent further damage to the environment and infrastructure.

Various techniques are employed to stabilize the burned landscape. These may include reseeding with native grasses, replanting trees and shrubs, and employing mechanical methods to control erosion. The implementation of BAER measures is paramount for safeguarding water quality, preventing landslides, and mitigating the risk of floods. BAER efforts are frequently integrated with other post-wildfire management activities to ensure a comprehensive approach to recovery.

7.3.4 ECOSYSTEM RESTORATION

Restoring vegetation is a key aspect of post-wildfire recovery (Anderson, 1995; Jackson and Hobbs, 2009). Replanting native species and reintroducing vegetation contribute to soil stabilization, enhanced water retention, and the restoration of wildlife habitats. Restoration efforts are tailored to the specific requirements of the burned area, taking into account soil conditions, climate, and ecological factors. For instance, the restoration of fish and wildlife habitats, including watersheds and instream environments, may be necessary as fire impacts can alter habitats for various species.

Wildfires can have a profound impact on wildlife habitats, causing displacement and the depletion of essential resources. Habitat restoration involves creating conditions that support wildlife and promote ecological balance (Miller and Hobbs, 2007). This may encompass planting vegetation, establishing wildlife corridors, and monitoring animal populations to ensure the successful recovery of the affected area. While fire can enhance habitat for certain species, it may have undesirable effects on others, including those that are already endangered or threatened, such as federally listed endangered or threatened species.

Post-wildfire landscapes can become susceptible to invasive species, which can outcompete native vegetation and further disrupt ecosystems (Johnson et al., 2006; Keeley, 2006). Effective post-fire restoration entails controlling and managing invasive species through both immediate and sustained actions. This involves reseeding and replanting with native species to restore ecological balance and implementing long-term controls to manage non-native species. Preventing invasive species from establishing dominance is paramount for the successful recovery of native ecosystems.

Continuous monitoring of restored areas is also paramount to evaluate the efficacy of restoration endeavors. Adaptive management approaches facilitate adjustments based on monitoring data and evolving environmental conditions. This ensures that restoration efforts remain effective and responsive to changing circumstances. The process can span several years and necessitates ongoing evaluation to address emerging challenges and refine strategies.

7.3.5 COMMUNITY SUPPORT

After a wildfire, providing emergency assistance and relief to affected communities is a priority (Edgeley and Paveglio, 2017). This encompasses providing temporary housing, financial assistance, and medical services. Collaborating with local organizations and agencies facilitates the efficient delivery of support and ensures that it meets the specific needs of the affected individuals.

Concurrently, restoring access to and repairing infrastructure is a crucial aspect of community support following a wildfire. This encompasses the restoration of roads, trails, bridges, and utilities, which are indispensable for daily life and economic recovery.

Repairing recreational infrastructure, including trailheads and campgrounds, is paramount for communities and businesses that rely on outdoor recreation. Efficient repair and reconstruction efforts mitigate further resource degradation, facilitate economic recovery, and restore normalcy for impacted residents.

Clear communication with the public is crucial for maintaining residents' informedness regarding recovery initiatives. Public information campaigns, community gatherings, and informative resources facilitate the dissemination of updates, address pertinent concerns, and enhance community resilience. Effective communication also helps in managing public expectations and facilitating community engagement in recovery efforts.

The psychological impact of wildfires can be significant (Finlay et al., 2012; Palinkas and Wong, 2020). Providing mental health support and counseling services assists individuals in coping with trauma and stress. Community-based support programs and access to mental health resources are crucial for overall recovery. Furthermore, addressing mental health needs is an integral component of supporting individuals and communities throughout the recovery process.

7.3.6 LONG-TERM RECOVERY PLANNING

Long-term recovery planning involves developing strategies for rebuilding and restoring communities and infrastructure (Bates and Peacock, 1989; Shaw, 2014). This encompasses the development of comprehensive recovery plans, securing funding, and collaborating with stakeholders to implement a holistic approach to reconstruction. Effective planning serves as a guiding principle for recovery efforts, addressing both immediate and long-term requirements.

Enhancing community resilience to future wildfires is a crucial aspect of long-term recovery. This encompasses improving building codes, establishing defensible spaces around properties, and promoting preparedness and education. By fostering resilience, communities can mitigate vulnerability to future wildfires and facilitate sustainable recovery. Developing and implementing strategies to enhance resilience is paramount in reducing future risks and bolstering recovery efforts.

Post-wildfire recovery also presents an opportunity to assess the efficacy of response and recovery measures. Lessons derived from previous wildfires serve as valuable inputs to enhance future preparedness and management strategies. This ongoing evaluation process contributes to the overall effectiveness of wildfire management and recovery endeavors.

7.4 TIME AND COST OF RESTORATION

Post-wildfire restoration is a protracted process that spans several years. The intricacy of restoring ecosystems, repairing infrastructure, and supporting impacted communities necessitates unwavering effort and coordinated collaboration. Restoration endeavors, encompassing revegetation, erosion control, and habitat restoration, necessitate substantial time to attain desired outcomes and ensure enduring recovery.

The financial consequences of post-wildfire restoration are significant. The cost of implementing erosion control measures, rebuilding infrastructure, and restoring ecosystems can be substantial. Ongoing maintenance and monitoring also contribute to the overall cost. Given the current high costs of wildfire suppression, which now consume over fifty percent of various forest services budgets, allocating sufficient resources for restoration efforts is challenging but essential.

7.5 DATA AND POST-FIRE MANAGEMENT

The integration of robust data collection and analysis into post-fire management practices is crucial for effective assessment, resource allocation, restoration planning, monitoring, risk reduction, and policy development. Data not only enhances the efficiency and effectiveness of recovery efforts but also supports the development of resilient ecosystems and communities in the aftermath of fire events. Let us explore the specific areas where data is essential in post-fire management.

7.5.1 ASSESSMENT OF FIRE IMPACT

Data collection through remote sensing technologies, such as satellite imagery and aerial photography, provides crucial insights into the spatial extent of burned areas. This information is pivotal for generating burn severity maps, which categorize the intensity of fire damage across different regions. This spatial analysis enables the precise quantification of affected landscapes and the identification of high-severity zones where immediate intervention may be necessary.

Field surveys and ecological assessments are employed to evaluate the impact of fire on various ecological components, including vegetation types, soil properties, and wildlife habitats. Detailed vegetation inventories and soil sampling provide data on post-fire recovery conditions and the extent of degradation. Furthermore, wildlife monitoring data, collected through camera traps and tracking methods, assists in assessing the direct and indirect effects of fire on animal populations and their habitats.

7.5.2 RESOURCE ALLOCATION AND MANAGEMENT

Data on fire severity and ecological impact facilitates the prioritization of restoration efforts. Areas with high burn severity and substantial ecological damage are designated as high-priority zones for immediate intervention. This prioritization is essential for directing resources, including personnel, financial support, and technical expertise, to where they are most urgently required.

Efficient resource management also necessitates data-driven decision-making. Geographic Information System (GIS) tools and data analytics are used to optimize the allocation of resources such as firefighting equipment, restoration materials, and human resources. By analyzing patterns and trends in fire damage, decision-makers can strategically deploy resources to maximize their effectiveness.

7.5.3 PLANNING AND IMPLEMENTATION OF RESTORATION STRATEGIES

Data on soil health, encompassing parameters such as soil moisture content, nutrient levels, and erosion susceptibility, serves as a foundation for the formulation of targeted restoration strategies through a comprehensive recovery needs assessment. For instance, soil erosion models can elucidate potential risks and subsequently guide the implementation of erosion control measures, such as silt fences or erosion barriers. Additionally, data on pre-fire vegetation conditions helps in selecting appropriate plant species for revegetation efforts.

The selection of plant species for replanting is also guided by data on soil conditions, local climate, and native plant communities. Seed banks and nursery inventories are utilized to select species that are resilient and adapted to the post-fire environment, often from those that are currently available. Restoration plans are developed based on data-driven models to ensure the successful establishment of vegetation and the recovery of wildlife habitats.

7.5.4 MONITORING AND EVALUATION

Continuous data collection is paramount for monitoring the progress of restoration activities. Key performance indicators, such as vegetation cover, species diversity, and soil stability, are meticulously monitored to assess the efficacy of restoration initiatives. This ongoing evaluation facilitates adaptive management, enabling strategies to be modified in response to real-time data to address emerging challenges.

Longitudinal studies, involving repeated data collection over extended periods, provide insights into the long-term recovery of ecosystems. These studies evaluate the resilience and sustainability of restored areas by assessing trends in ecological variables, including biodiversity and ecosystem function. Long-term data enables the understanding of ecological recovery trajectories and informs future fire management practices.

7.5.5 RISK REDUCTION AND PREPAREDNESS

Data on historical fire patterns, coupled with climate models and vegetation maps, facilitates the prediction of future fire risks. This predictive modeling enables the identification of high-risk areas, prompting the implementation of preventive measures such as controlled burns and vegetation management to mitigate potential fire hazards.

Concurrently, data on the socioeconomic impact of fires on local communities supports the development of community preparedness plans. Surveys and impact assessments provide information on displacement, economic losses, and recovery requirements. This data is utilized to enhance community resilience through targeted support programs, public education initiatives, and infrastructure enhancements.

7.5.6 POLICY AND DECISION-MAKING

Data-driven insights are pivotal in shaping fire management policies and regulations. Empirical evidence on fire behavior, ecological impacts, and restoration outcomes informs policy decisions pertaining to land use, conservation priorities, and fire management practices. Data supports the development of evidence-based policies that address both immediate recovery requirements and long-term sustainability objectives.

Comprehensive data on fire impacts and recovery requirements also serves as a foundation for advocating for funding and resources. Detailed reports and analyses present a compelling case for financial support and policy attention, ensuring that adequate resources are allocated to support effective post-fire management and recovery efforts.

7.6 CASE STUDY: YELLOWSTONE NATIONAL PARK WILDFIRE OF 1988

Yellowstone National Park, established in 1872 as the world's first national park, stretches across the states of Wyoming, Montana, and Idaho (Haines, 1974; Jackson, 1942). Covering approximately 3,472 square miles (8,991 square kilometers), it is renowned for its geothermal features, diverse ecosystems, and rich biodiversity (Chittenden, 2018; Keefer, 1971). The park's landscape encompasses geothermal hot springs, mountain ranges, and extensive forests, providing habitat for renowned species such as grizzly bears, wolves, elk, and bison. Yellowstone National Park draws millions of visitors annually due to its unparalleled natural beauty and scientific importance.

Historically, fire management in Yellowstone focused primarily on suppression, driven by a belief that fire was inherently destructive to both human interests and natural resources (Barker, 2013). This approach was implemented to safeguard the park's infrastructure and ensure the safety of visitors. However, by the 1970s and 1980s, there was a significant shift in understanding the role of fire in ecosystems (Sellers and Despain, 1976; Taylor, 1973). Acknowledging the ecological advantages of fire, including its contribution to nutrient cycling, habitat formation, and biodiversity preservation, prompted the adoption of more nuanced fire management approaches. These strategies sought to balance fire suppression with allowing some natural fires to burn, reflecting a growing appreciation for fire's role in maintaining healthy ecosystems (Barker, 2013).

7.6.1 THE 1988 FIRE EVENT

The 1988 wildfire season in Yellowstone is one of the most notable fire events in the park's history and a watershed moment in wildfire management (Romme and Despain, 1989). The wildfires commenced in early June 1988, with initial ignitions attributed to a confluence of human activities and lightning strikes. By mid-July, a series of lightning storms further ignited additional fires, accelerating their rapid spread. By late August, the scale of the fires had reached unprecedented levels, burning over 800,000 acres — about 36% of the park's total area [Figure 2.2(NASA, 2019). The colossal scale and intensity of the fires were clearly evident through satellite imagery and extensive media coverage, underscoring the gravity of the situation.

The fires were complex, with multiple burn fronts and varying intensities (Turner et al., 2003). Hot, dry, and windy conditions further exacerbated the situation, making containment efforts particularly challenging (van Wagtendonk, 1991). The extensive and intense nature of these fires not only altered the park's landscape but also posed substantial risks to nearby communities and infrastructure. The fires' scale and complexity necessitated a critical evaluation of fire management strategies and resource allocation.

7.6.2 RESPONSE AND MANAGEMENT

The initial response to the fires involved the deployment of extensive firefighting resources, including personnel, aircraft, and equipment. However, the fires' magnitude and intensity soon surpassed the capabilities of traditional firefighting methods. By August 1988, it became apparent that a new approach was necessary (Christensen et al., 1989; Elfring, 1989). The National Park Service, together with other federal agencies, adopted a strategy of "let burn" for certain fires (Barker, 2013). This strategy permitted some fires to burn naturally, acknowledging their role in ecological processes, while focusing resources on protecting critical infrastructure and key ecological areas.

Several challenges complicated the response to the 1988 fires (Turner et al., 2003). The magnitude and intensity of the wildfires strained firefighting resources to their limits. With multiple fires burning concurrently, controlling and containing the blazes proved challenging. Adverse weather conditions, such as hot, dry, and windy weather, further exacerbated the fires, complicating containment operations. The fires' impact on diverse habitats, including critical wildlife areas and old-growth forests, highlighted the need for a more nuanced approach to fire management that balanced immediate response with long-term ecological considerations.

7.6.3 RECOVERY AND LONG-TERM IMPACT

Following the wildfires, Yellowstone underwent substantial ecological and environmental transformations. Extensive forest areas were reduced to charred landscapes, resulting in increased soil erosion and altered water flow patterns. Wildlife populations faced habitat loss and displacement, adding to the complexity of recovery efforts (Minshall et al., 1989; Pearson et al., 1995; Turner et al., 1994).

More than 30 years later, many areas within Yellowstone National Park have still not fully recovered from the 1988 fires [Figure 2.2(Li et al., 2022; NASA, 2019; Romme and Turner, 2004; Romme et al., 2011). The regrowth of vegetation in certain regions has been gradual and incomplete. High-severity burn areas, in particular, continue to exhibit sparse vegetation and limited recovery. The intense fires disrupted natural regeneration processes, resulting in ongoing challenges for vegetation recovery.

Soil erosion persists as a continual challenge in numerous regions of the park. The depletion of vegetation, particularly in areas subjected to severe burns, has perpetuated erosion challenges that adversely impact water quality and sedimentation patterns within streams and rivers. These ongoing issues underscore the enduring effects of the fires on the park's hydrological processes and ecosystem dynamics.

Wildlife habitats altered by the fires have had enduring effects. Some species have adapted to the altered conditions, while others continue to encounter challenges due to habitat fragmentation and alterations in food availability. The park's wildlife populations have had to adjust to the transformed landscape, with varying degrees of success.

The post-fire environment has also witnessed the proliferation of invasive plant species. These invasive species possess the ability to outcompete native vegetation, thereby further complicating the recovery of native plant communities and disrupting ecosystem dynamics.

Ongoing monitoring and management are essential for addressing the long-term impacts of the 1988 fires. Ecological monitoring continues to assess changes in vegetation, soil conditions, and wildlife populations, providing valuable data for evaluating recovery progress. Restoration projects have been implemented to address specific issues, such as soil erosion and invasive species control, in order to support the park's ecological health and facilitate recovery.

Furthermore, adaptive management strategies have been implemented to address emerging challenges and evolving conditions. These strategies prioritize flexibility and responsiveness, enabling park management to adjust their approaches in accordance with the changing requirements of the ecosystem.

7.6.4 CONCLUSION

The 1988 Yellowstone National Park wildfire stands as a pivotal moment in the annals of wildfire management history. Not only did it expose the immediate challenges associated with managing large-scale wildfires, but it also catalyzed a paradigm shift in fire management practices. The consequences of adoption of the "let burn" strategy and the recognition of fire's long-term ecological significance marked a profound transformation in the approach to fire management within protected areas.

Over three decades after the devastating wildfires, Yellowstone National Park continues to grapple with the long-term consequences of these events. The ongoing vegetation recovery, persistent erosion, and habitat alterations serve as stark reminders of the profound impact of wildfires. This case study serves as a valuable lesson in fire management, highlighting the imperative for continuous monitoring, targeted restoration efforts, and adaptive management approaches. It underscores the intricate nature of ecological recovery following large-scale disturbances and emphasizes the importance of integrating ecological principles into fire management strategies.

7.7 CASE STUDY: BANDHAVGARH TIGER RESERVE WILDFIRE OF 2021

Bandhavgarh Tiger Reserve, located in the Umaria district of Madhya Pradesh, India, is renowned for its rich biodiversity and significant conservation value (Kishnani, 2019). Spanning an area of approximately 1,536 square kilometers, the reserve is divided into three distinct zones: the core area, the buffer zone, and the surrounding forested landscape.

Bandhavgarh National Park was established in 1968 and subsequently designated as a tiger reserve in 1993. The Tiger Reserve is home to a diverse range of flora and fauna, with the Bengal tiger being the primary species of interest (Chandra, 2006; Dubey et al., 2021).

The diverse topography of Bandhavgarh Tiger Reserve, characterized by hills, valleys, and dense forests, sustains a rich and intricate ecosystem. The vegetation types within the reserve encompass a spectrum, ranging from tropical dry and moist deciduous forests to sal (*Shorea robusta*) stands and bamboo forests.

This ecological diversity provides vital habitats for a multitude of species, such as the Royal Bengal tiger, leopards, wild boars, sambar deer, and a diverse array of bird species. The reserve also functions as a pivotal corridor for wildlife movement between other protected areas within the region. All of these are threatened by wildfires (Pati et al., 2024).

Bandhavgarh Tiger Reserve has been a cornerstone of conservation efforts in India for an extended period. The reserve's management strategy emphasizes the protection of tiger populations and their habitats, anti-poaching measures, and habitat restoration (Nath, 2000). The presence of a well-established network of anti-poaching units and active monitoring programs contributes to the reserve's role as a critical sanctuary for endangered species (Chouksey and Singh, 2018). Furthermore, Bandhavgarh engages in eco-tourism activities that promote conservation

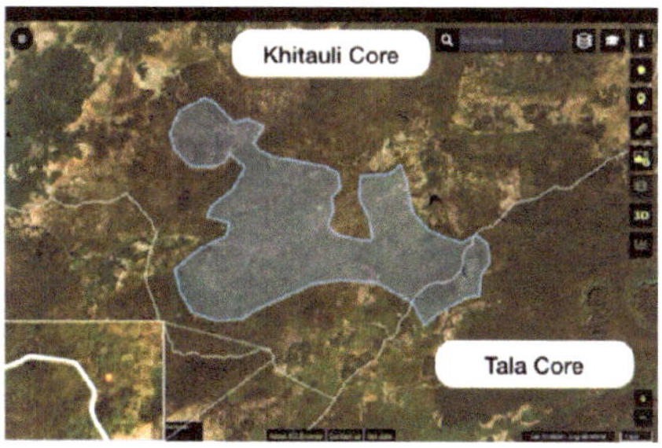

(a) In this satellite image from 14 March 2021, fires are visible outside the area of interest.

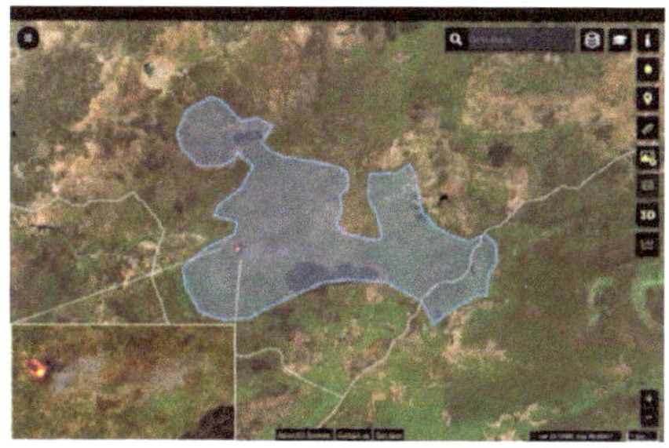

(b) In this satellite image from 29 March 2021, the fire inside the area of interest is visible, along with a burn scar from a previous fire.

(c) In this satellite image from 03 April 2021, we find that the area of interest is completely burnt.

(d) In this satellite image from 10 October 2021, we find that the area of interest has turned green.

FIGURE 7.2: Bandhavgarh wildfire event of 2021, visualized in EO browser, https://apps.sentinel-hub.com/eo-browser/.

awareness and generate revenue to support ongoing conservation initiatives (Kishnani, 2019). Consequently, the Tiger Reserve holds immense significance for the region.

7.7.1 THE WILDFIRE EVENT

In 2021, Bandhavgarh Tiger Reserve experienced a catastrophic wildfire that began in the second week of March [Figure 7.2] (Pati et al., 2024). The origin of the fire is believed to be human-caused burning of agricultural and leaf residues in the surrounding areas, a prevalent practice that inadvertently leads to the spread of fire into forested regions. Early detection of the fire through satellite imagery proved pivotal, enabling timely response and intervention.

By late March, the severity of the wildfire became increasingly evident. Satellite images [Figure 7.2b–c] revealed extensive areas of scorched forest. The fire's intensity resulted in significant ecological damage, including the destruction of critical habitats and disruption of wildlife (Banerjee et al., 2024).

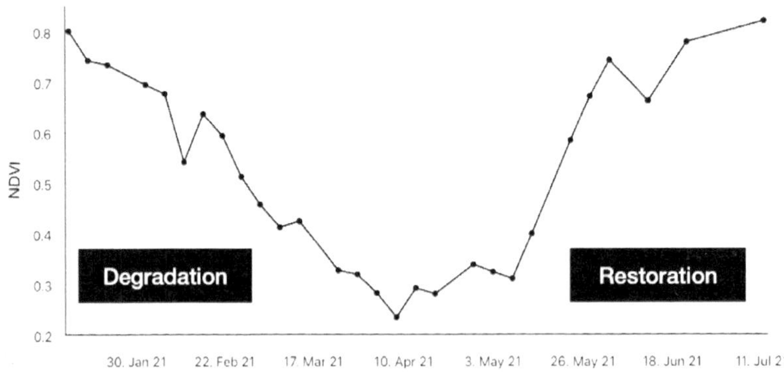

FIGURE 7.3: Recovery process in the Bandhavgarh wildfire event of 2021.

7.7.2 RESPONSE AND CHALLENGES

Despite the deployment of substantial resources and personnel, the fire's rapid spread and intensity presented formidable challenges. By early April, a considerable portion of the reserve's dense forest had been consumed by the flames [Figure 7.2c]. The extensive damage surpassed initial containment efforts, necessitating a reassessment of strategies and fostering increased collaboration with local communities.

The involvement of local samiti (community) members was a pivotal juncture in the restoration process. These community groups, possessing local expertise and a vested interest in the reserve, played a critical role in supporting recovery endeavors. This encompassed replanting native species, facilitating natural growth through assisted natural regeneration, and promptly removing any emerging invasive species. By the commencement of October, their contributions, in conjunction with administrative efforts, culminated in a remarkable restoration of the affected areas. Satellite images from this period demonstrated a significant recovery, with minimal visible signs of the previous damage [Figure 7.2d]. Of course, the long-term effects remain to be seen in the future.

7.7.3 RESTORATION AND RECOVERY

The recovery of Bandhavgarh Tiger Reserve can be assessed through the NDVI (Normalized Difference Vegetation Index) plot [Figure 7.3] (Singh and Jeganathan, 2024). The NDVI data, which measures vegetation health and density, showed a sharp decline immediately following the fire, indicative of the loss of foliage. However, the subsequent rebound in the NDVI curve illustrated the resilience of the forest ecosystem and the effectiveness of restoration efforts.

Restoration efforts employed a combination of natural regeneration and human-assisted techniques. Natural regeneration was further facilitated by seasonal rainfall, which stimulated the growth of new vegetation. Concurrently, assisted natural regeneration and artificial reforestation were implemented, involving the planting of native saplings and the implementation of soil and water conservation measures.

The local samitis (communities) played a pivotal role in the restoration process. Their participation in pre-plantation, plantation, and post-plantation activities provided crucial support for the recovery efforts. Not only did this offer employment opportunities, but the community also benefited from a share of the revenue generated from forest resources. This collaborative approach not only enhanced the effectiveness of the restoration activities but also fostered a sustainable approach to forest management.

It is noteworthy that NDVI quantifies the extent of vegetation, not the biomass content of the area. Re-emergence of greenness is often faster in the Central Indian landscape because most of the wildfires are only surface fires, and many of the fire-resistant trees, adapted to surface fires, are able to survive these fires. A comparable situation exists with other plants that possess the ability to survive in the form of rootstock or fire-resistant seeds. Nevertheless, while it is easy for lots of vegetation to come up quickly after a wildfire, their growth to maturity does take a very long time. Only after the plants have fully matured will they resume their previous ecological roles following the wildfire. Nevertheless, the newly emerged plants will require extensive care and substantial resources for an extended period. It is also important to note that NDVI does not measure changes and recovery in the populations of animals, birds, insects, microorganisms, and other forms of life. Thus, the augmentation of NDVI, while encouraging, does not mean that the restoration is complete.

7.7.4 CONCLUSION

The Bandhavgarh Tiger Reserve wildfire of 2021 serves as a significant case study in the context of forest fire management and ecological restoration. The management operations, conceptualized with the aid of the knowledge gleaned from the Yellowstone National Park wildfires of 1988, tried to put a quick stop to the wildfires. Thus, while the initial devastation was profound, the combined efforts of forest staff, local communities, and conservationists led to a quick start of the recovery process. This case emphasizes the paramount importance of integrated fire management strategies, early detection mechanisms, community engagement, and adaptable restoration techniques. In a manner, Bandhavgarh exemplifies the resilience of ecosystems and the potential for effective post-fire management, providing hope and valuable lessons for future conservation endeavors.

Section III

Technologies & best practices

8 Remote Sensing

Human impact on ecosystems and the environment has been growing over time, as shown by the equation (Awadhiya, 2021):

$$I = P \times A \times T$$

In this equation, I stands for the impact humans have on the environment, P is the human population (or population density if we're considering impact per unit area), A represents human affluence (the amount of resources we consume), and T indicates the level of technology that helps us shape our surroundings and extract resources from our environment.

It is clear that human impact increases with the increasing number of people, growing resource needs, and advancement of technology. Over time, all three of these factors have risen: our population has grown, our demand for resources has increased as we have become wealthier, and our technology has advanced, allowing us to extract more resources more quickly and more efficiently. And so our impact on our environment has also ballooned over time.

One significant consequence of these growing impacts is the increase in wildfires. Most wildfires around the world are caused by human activities (Robinne and Secretariat, 2021). Some wildfires are intentionally set, such as to drive animals into traps, clear land for farming, or divert authorities from crime scenes. Others are more accidental and less intentional, including fires spreading from campsites or agricultural fields, sparks from electrical lines, or even the effects of climate change. As the global population expands and our resource demands increase, the frequency and intensity of human-caused fires, both intentional and accidental, also rise.

While we can't quickly reduce the human population or stop progress, we can use technology to address these issues. The problems created by technology can often be solved with more technology. Green technologies, for example, help us reduce environmental impacts while still enjoying the benefits of development. Similarly, new technologies can help manage and reduce wildfires. Improved firefighting techniques and better resource allocation can make our efforts more effective. This chapter, along with the next ones, will explore some of these innovative technologies and their potential benefits.

8.1 FUNDAMENTALS

In the previous chapters, we have seen how the use of satellite imagery has revolutionized wildfire management. Satellites provide repeated measurements in a cost-effective manner, even in challenging terrains and inaccessible areas. Over time, satellite imagery has enabled us to identify trends, forecast, and model wildfires. This capability falls under the realm of 'remote sensing,' defined as sensing from a distance. Let us now delve deeper into remote sensing.

8.1.1 WHAT IS REMOTE SENSING?

Remote sensing is the process of acquiring information about objects or phenomena without directly being in physical contact with them (Awadhiya, 2016). The term 'remote' refers to the distance from which the information is gathered, and 'sensing' pertains to the process of collecting data about the target.

Remote sensing comprises three fundamental components:

1. **Target object:** The entity or phenomenon whose information is to be acquired.
2. **Sensor:** A device capable of detecting and recording data from a distance, typically storing it for subsequent analysis.
3. **Transmission method:** The mechanism through which information is transmitted from the target to the sensor.

Remote sensing is a widely utilized technology in both natural and technological domains. Within the biological realm, numerous organisms, including humans, rely on remote sensing to perceive and respond to their surroundings. For instance, animals may detect a predator through the rustling of leaves or locate a branch to jump without direct physical contact with either the predator or the branch. Similarly, our sensory organs are designed to gather information from a distance. When we perceive a friend approaching, we utilize light waves to discern their presence without physical contact. In this scenario, the target is the friend, the sensor is the eye, and the transmission medium is light waves. Similarly, when we hear a bee buzzing, sound waves convey information to our ears, enabling us to detect the bee's presence without physical contact. In this instance, the target is the bee, the sensor is the ear, and the transmission medium is sound waves.

Through technological advancements, we have developed numerous remote sensors. For instance, a microphone functions similarly to our ears, capturing sound. A camera mimics the eye, with its diaphragm resembling the iris and its lens akin to the eye's lens. A camera's film or electronic sensor parallels the retina. Satellite sensors operate on analogous principles to a camera, albeit designed to gather information from space. While they cannot detect sound waves (as sound necessitates a medium for transmission), they utilize electromagnetic waves, which can traverse the vacuum of space.

8.1.2 ELECTROMAGNETIC SPECTRUM

Satellite remote sensing utilizes electromagnetic waves to transmit information from the target to the sensor. Electromagnetic waves are generated when an electric field oscillates in conjunction with a magnetic field, resulting in a wave that carries energy perpendicular to both fields. Unlike sound waves, electromagnetic waves do not necessitate a medium and can traverse a vacuum.

Electromagnetic waves are characterized by their wavelength (λ) and frequency (ν). Wavelength is the distance over which the wave's shape repeats, or the distance between corresponding points of consecutive waves. It is measured in units of length such as meters, centimeters, or millimeters.

Frequency is the number of waves that pass a given point in a fixed period of time. It is measured in units of per time, with the SI unit being hertz (Hz). One hertz equals one event per second, or s^{-1} in SI base units.

The relationship between wavelength, frequency, and the speed of light in a vacuum (c) — defined as exactly 299,792,458 meters per second — is given by:

$$c = \lambda\nu$$

Rearranging this equation yields:

$$\lambda = \frac{c}{\nu}$$

$$\nu = \frac{c}{\lambda}$$

Thus, wavelength and frequency are inversely related — as one increases, the other decreases.

When we look at Sun's light through a prism, we can observe a spectrum where differently colored lights get separated based on their wavelengths. The electromagnetic spectrum includes this, and also other wavelengths that we cannot see through our eyes. It is categorized based on wavelength into various types of waves, including radio waves, microwaves, infrared, optical light, ultraviolet light, X-rays, and gamma rays [Table 8.1](Center, 2013b). Each type of wave has unique properties and applications, generally related to its energy and its ability to penetrate different materials.

The energy E of a photon in an electromagnetic wave is governed by the equation:

$$E = h\nu = \frac{hc}{\lambda}$$

where h denotes Planck's constant ($6.62607015 \times 10^{-34}$ Joule-second), ν is the frequency of the wave, c is the speed of light in a vacuum ($299,792,458$ meters per second), and λ represents the wavelength. This relationship shows that photon energy is directly proportional to the frequency of the electromagnetic wave and inversely proportional to its wavelength. As the frequency increases (or the wavelength

TABLE 8.1
Wavelength, frequency, and energy in the electromagnetic spectrum

Region	Wavelength (m)	Frequency (Hz)	Energy (J)
Radio	$> 1 \times 10^{-1}$	$< 3 \times 10^{9}$	$< 2 \times 10^{-24}$
Microwave	$1 \times 10^{-3} - 1 \times 10^{-1}$	$3 \times 10^{9} - 3 \times 10^{11}$	$2 \times 10^{-24} - 2 \times 10^{-22}$
Infrared	$7 \times 10^{-7} - 1 \times 10^{-3}$	$3 \times 10^{11} - 4 \times 10^{14}$	$2 \times 10^{-22} - 3 \times 10^{-19}$
Optical	$4 \times 10^{-7} - 7 \times 10^{-7}$	$4 \times 10^{14} - 7.5 \times 10^{14}$	$3 \times 10^{-19} - 5 \times 10^{-19}$
UV	$1 \times 10^{-8} - 4 \times 10^{-7}$	$7.5 \times 10^{14} - 3 \times 10^{16}$	$5 \times 10^{-19} - 2 \times 10^{-17}$
X-ray	$1 \times 10^{-11} - 1 \times 10^{-8}$	$3 \times 10^{16} - 3 \times 10^{19}$	$2 \times 10^{-17} - 2 \times 10^{-14}$
Gamma-ray	$< 1 \times 10^{-11}$	$> 3 \times 10^{19}$	$> 2 \times 10^{-14}$

decreases), the energy of the photon increases correspondingly. This principle is crucial in understanding the behavior of different types of electromagnetic waves across the spectrum.

8.1.3 INTERACTION OF WAVES WITH MATTER

Electromagnetic waves interact with materials through several fundamental processes. These interactions play a crucial role in determining how we perceive objects and how sensors detect these waves:

1. **Reflection**: Reflection is a phenomenon that occurs when electromagnetic waves encounter a surface and are redirected back into the medium from which they originated. The fundamental principle governing reflection is the law of reflection, which establishes that the angle of incidence (the angle formed between the incoming wave and the normal to the surface) is equal to the angle of reflection (the angle formed between the reflected wave and the normal). Reflection plays a crucial role in the visibility of objects and is an essential component of optical devices such as mirrors and telescopes.

2. **Absorption**: Absorption is the process through which the energy of an electromagnetic wave is absorbed by the material it encounters. This energy is frequently converted into heat or other forms of energy. The extent of absorption is influenced by the material's properties, including its color, molecular composition, and electronic structure. Black surfaces exhibit greater absorption of visible light compared to white or reflective surfaces. Absorption plays a pivotal role in applications such as spectroscopy, wherein the material's interaction with specific wavelengths provides insights into its composition and temperature.

3. **Refraction**: Refraction involves the bending of electromagnetic waves as

they pass through materials of different optical densities. When waves enter a medium with a different optical density, their speed changes, causing the waves to bend. This bending is described by Snell's law, which relates the angles of incidence and refraction to the indices of refraction of the two media. Refraction is responsible for phenomena such as the distortion of objects under water and the focusing of light in lenses and microscopes.

4. **Transmission**: Transmission is the passage of electromagnetic waves through a medium. The transmission capability of a medium for electromagnetic waves is contingent upon its transparency and the specific wavelengths involved. For instance, glass exhibits transparency to visible light but opacity to ultraviolet and infrared radiation. Transmission is paramount for communication technologies such as optical fibers and radio waves, which rely on the ability of materials to convey signals over extended distances. Furthermore, the transmission of electromagnetic waves through the atmosphere (and even through a vacuum) is the enabler of satellite remote sensing.

8.1.4 ATMOSPHERIC INTERACTIONS AND ATMOSPHERIC WINDOW

Electromagnetic waves traversing the Earth's atmosphere undergo intricate interactions. The atmosphere selectively absorbs and scatters specific wavelengths of radiation, permitting others to pass through. This phenomenon results in the formation of the "atmospheric window," a range of wavelengths minimally impacted by atmospheric interference. This window holds paramount importance for remote sensing applications, as it delineates the wavelengths suitable for observing the Earth's surface from space or aerial platforms.

The atmospheric window enables electromagnetic waves to penetrate the Earth's atmosphere and return to the sensor with minimal distortion. This phenomenon is crucial for accurate data collection and interpretation in remote sensing applications. Variations within this electromagnetic wave spectrum can provide valuable insights into surface properties, including temperature, vegetation cover, and urban development.

8.1.5 PASSIVE AND ACTIVE REMOTE SENSING

Remote sensing relies on the generation, detection, and analysis of electromagnetic waves. Based on the source of electromagnetic radiation, remote sensing is classified into two primary categories — passive and active remote sensing.

1. **Passive Remote Sensing**: Passive remote sensing employs natural sources of electromagnetic radiation, such as sunlight. The Sun's energy traverses the vacuum of space and the Earth's atmosphere, ultimately reaching the Earth's surface and interacting with it. The reflected or emitted radiation

from the surface subsequently traverses back through the atmosphere to a sensor, such as an eye, photographic plate, or satellite sensor. This process is akin to observing an object during daylight using the Sun's energy. Instruments for passive remote sensing are typically simpler and less expensive because they do not necessitate an active source of radiation. Notable examples include the Landsat satellites operated by NASA and the Sentinel-2 series of satellites operated by the European Space Agency. These instruments rely on the inherent variations in reflected sunlight to provide data on land cover, vegetation health, and other surface characteristics.

2. **Active Remote Sensing**: Active remote sensing involves the emission of electromagnetic radiation by the sensor, which subsequently interacts with the target and is returned to the sensor. This process is analogous to using a flashlight to illuminate an object in darkness. Active remote sensing systems necessitate a power source to generate the radiation and are generally more intricate and costly. Notable examples include NASA's GEDI LiDAR mission, which measures forest canopy heights, and the European Space Agency's Sentinel-1 RADAR mission, which provides detailed information on surface deformations and land cover changes. Active remote sensing enables detailed observations even in low-light or no-light conditions or through obscured or cloudy atmospheres (when employing non-optical wavelengths such as radio waves). Consequently, it holds significant value in applications such as environmental monitoring, disaster management, and infrastructure assessment.

Both passive and active remote sensing play pivotal roles in data collection and environmental monitoring. Each method possesses distinct advantages and is selected based on the specific requirements of observation and the characteristics of the target area.

8.2 EARTH-OBSERVATION SATELLITES

Earth-observation satellites are advanced spacecraft positioned in orbit around our planet to provide continuous and comprehensive monitoring. These satellites are instrumental in a range of applications including weather forecasting, environmental monitoring, resource management, cartography, and even military surveillance. Their ability to observe and collect data from space has revolutionized our understanding of Earth and its processes.

8.2.1 HISTORICAL DEVELOPMENT

The concept of capturing images of Earth from above has evolved significantly over the centuries. Early attempts at aerial photography included the use of balloons, which, in the mid-19th century, allowed for the collection of images from elevated positions. By the early 20th century, pigeon cameras, which were mounted on pigeons, provided a novel approach to capturing aerial imagery. During the World Wars, aircraft-mounted cameras played a critical role in reconnaissance, offering a strategic advantage. The dawn of space-based remote sensing began in 1946 with a sub-orbital V-2 rocket equipped with a camera. This marked the first step towards the extensive satellite imagery we rely on today. The Landsat program, initiated in 1972, represents a significant milestone in Earth observation, providing the first systematic satellite imagery.

8.2.2 SATELLITE ORBITS

Satellites are positioned at various altitudes, which impacts their observational capabilities and operational characteristics. Satellites in low Earth orbit (LEO) operate at altitudes ranging from about 200 km to 2,000 km. These satellites provide high-resolution images and cover small areas in great detail. However, they experience significant atmospheric drag, which requires more fuel for orbit maintenance and regular re-boost operations. Conversely, satellites in higher orbits, such as those in a geostationary orbit, can observe larger portions of the Earth but with less detail. These higher orbits, located at approximately 35,786 km above the Earth, require substantial energy for launch and orbital insertion. Geostationary satellites appear fixed relative to a specific point on the Earth's surface, which is ideal for continuous monitoring of weather patterns, fire spread, and communications.

8.2.3 TIME PERIOD OF A SATELLITE

Satellites typically move in elliptical orbits. However, to better understand satellite orbits, we can simplify the scenario by considering a satellite moving in a circular orbit. Let m represent the mass of the satellite, and r its distance from the center of the Earth. The gravitational force acting on the satellite is given by:

$$F = \frac{G \times M \times m}{r^2}$$

where G is the gravitational constant (6.674×10^{-11} m^{3}kg^{-1}s^{-2}), and M is the mass of the Earth. This gravitational force provides the centripetal acceleration necessary for the satellite's circular motion. Thus,

$$\frac{G \times M \times m}{r^2} = m \times \frac{v^2}{r}$$

where v is the satellite's linear speed. Solving for v, we get

$$v = \sqrt{\frac{G \times M}{r}}$$

During one complete orbit, the satellite travels a distance of $2\pi r$. Therefore, its speed v can be expressed as:

$$v = \frac{2\pi r}{T}$$

where T is the orbital period. Rearranging to solve for T:

$$T = \frac{2\pi r}{v} = 2\pi\sqrt{\frac{r^3}{G \times M}}$$

This relationship indicates that as the distance r from the center of the Earth increases, the orbital period T also increases. Satellites in higher orbits take longer to complete a single orbit compared to those in lower orbits.

To compute the height of a geostationary satellite, we set T as 24 hours, to get an orbital radius of 42,164 kilometers. Subtracting the radius of the Earth, 6,378 kilometers gives the altitude of 35,786 kilometers relative to the Earth's surface.

8.2.4 TYPES OF SATELLITE ORBITS

Earth-observation satellites typically operate in one of two primary types of orbits, each serving specific observational needs:

1. **Sun-Synchronous Orbit**: Also known as a heliosynchronous orbit, this is a near-polar orbit where the satellite passes over the same point on Earth at the same local solar time each day. This consistency allows for observations under similar lighting conditions, which is valuable for monitoring changes over time. Sun-synchronous orbits are typically located at altitudes between 700 and 800 km, resulting in an orbital period of about 100 minutes.
2. **Geostationary Orbit**: Satellites in this orbit are positioned 35,786 km above the Earth's equator, where their orbital period matches the Earth's rotational period. This synchronization allows the satellite to remain fixed relative to a specific point on the Earth's surface. Geostationary orbits are particularly advantageous for continuous monitoring of weather patterns, communications, and broadcasting.

8.2.5 SATELLITE DATA

Satellite data is collected and stored in digital raster formats. A raster file represents data in a grid format, with information stored in a matrix of rows and columns. Each cell in this matrix is known as a pixel, and each pixel has a value called its digital number (DN). The DN represents the information captured by the pixel, and can include:

- **Photographic images**: RGB values from a photograph or scanned map.
- **Continuous data**: Measurements such as elevation, temperature, or spectral information collected by remote sensors.

- **Discrete data**: Categorical information such as land use classification or vegetation types.

Raster images are characterized by several key resolution parameters:

1. **Spatial resolution**: This refers to the sensor's capacity to discern between closely proximate objects on the Earth's surface. Enhanced spatial resolution facilitates the capture of finer details and more precise identification of features. For instance, a high spatial resolution may enable the detection of individual trees or even leaves, whereas lower resolution may only reveal broader forested areas. Factors that influence spatial resolution encompass the sensor's optical characteristics, its distance from the target, and atmospheric conditions. Higher spatial resolution yields more detailed images but also entails larger file sizes.

2. **Spectral resolution**: This indicates the sensor's capability to discern and record diverse electromagnetic radiation bands. For instance, a monochrome camera captures only brightness values across a broad spectrum, rendering various colors as the same shade of gray. Incorporating color filters enables the detection of specific wavelengths, resulting in images with distinct color bands. Higher spectral resolution entails employing additional filters, thereby capturing more detailed spectral information at the expense of larger file sizes.

3. **Radiometric resolution**: This describes the sensor's ability to discern differences in radiance or brightness levels. For example, a sensor with a 1-bit radiometric resolution might classify pixels as either black (0) or white (1), while a 2-bit resolution can differentiate four levels of brightness (00, 01, 10, and 11). Modern sensors typically use 8-bit or higher radiometric resolution, allowing for a greater range of brightness levels at the cost of larger file sizes.

4. **Temporal resolution**: This refers to the frequency at which a satellite revisits the same area of the Earth's surface. Higher temporal resolution enables more frequent observations, facilitating the detection of rapid temporal changes. However, this increased frequency necessitates a substantial volume of data storage and processing. Temporal resolution for an orbit can be enhanced by deploying multiple satellites within the same orbital configuration.

As an example, Landsat 7 satellite captures images with eight spectral bands, in spatial resolutions ranging from 15 to 60 meters, a temporal resolution of 16 days, and an 8-bit radiometric resolution. The detailed band information for Landsat 7 is summarized in table 8.2 (Center, 2013a).

TABLE 8.2
Band information for Landsat 7

Band	Wavelength (μm)	Resolution (m)
Band 1 - Blue	0.45-0.52	30
Band 2 - Green	0.52-0.60	30
Band 3 - Red	0.63-0.69	30
Band 4 - Near Infrared (NIR)	0.77-0.90	30
Band 5 - Shortwave Infrared (SWIR) 1	1.55-1.75	30
Band 6 - Thermal	10.40-12.50	60
Band 7 - Shortwave Infrared (SWIR) 2	2.09-2.35	30
Band 8 - Panchromatic	0.52-0.90	15

8.3 RADAR

RADAR, an acronym for "RAdio Detection And Ranging," is a sophisticated active remote sensing technology that employs radio waves to detect and analyze objects. It entails emitting radio waves towards a target. Upon encountering the target, a portion of the signal is reflected back to the RADAR system, which subsequently detects and processes this return signal. The elapsed time between the emission and reception of the signal, coupled with any variations in the signal's properties, furnishes comprehensive information regarding the target's distance, size, shape, and other pertinent characteristics.

In Section 8.1.4, we explored the concept of the atmospheric window, which refers to the range of wavelengths that can pass through the Earth's atmosphere with minimal absorption and scattering. RADAR technology significantly benefits from this concept. Unlike optical sensors that rely on visible light and can be severely impacted by atmospheric conditions such as clouds, rain, or fog, RADAR systems operate effectively unhindered across a broad range of microwave and radio frequencies. These frequencies experience minimal attenuation as they travel through the atmosphere, making RADAR particularly advantageous for continuous monitoring under adverse weather conditions. This capability is especially crucial for applications where persistent cloud cover is common, such as in certain rainforests and maritime / polar regions.

Furthermore, while passive optical remote sensing systems dependent on sunlight are rendered ineffective during the nighttime, RADAR systems can function continuously. This capability enables the acquisition of continuous data irrespective of the time of day, providing a significant advantage for monitoring and managing natural resources, environmental changes, and other phenomena that require regular observations.

8.3.1 SAR

Synthetic Aperture Radar (SAR) is an advanced form of RADAR technology that significantly enhances spatial resolution. As discussed in Section 8.2.5, high spatial resolution is crucial for capturing fine details in remote sensing imagery. SAR achieves this through a technique called synthetic aperture, which circumvents the limitations of the physical antenna size that can be deployed on satellites or aircraft.

In traditional RADAR systems, spatial resolution is directly proportional to the antenna's size: larger antennas yield finer resolution. However, practical constraints dictate the size of antennas that can be launched and maintained on satellites, spacecraft, and aircraft. To address this, SAR employs a method where a small antenna is utilized to collect data over the Earth. Each pass generates a segment of data for a specific ground area. By combining these segments computationally, SAR simulates the effect of a much larger antenna, effectively increasing the spatial resolution of the resulting imagery. This approach enables SAR to produce high-resolution images that reveal intricate features of the Earth's surface.

SAR operates across various frequency bands, each with distinct characteristics and applications. These bands are categorized based on their wavelength, and their detailed descriptions are provided in Table 8.3, courtesy of NASA Earth Data[1]. The selection of band influences the penetration depth, resolution, and the types of surfaces or materials that can be effectively analyzed. For instance, different bands may be more suitable for detecting vegetation, soil moisture, or urban infrastructure, and thus a combination of bands may enhance the versatility of SAR in environmental and scientific applications.

TABLE 8.3
Band information for SAR

Band	Frequency	Wavelength	Typical Application
Ka	27-40 GHz	1.1-0.8 cm	Airport surveillance
K	18-27 GHz	1.7-1.1 cm	Short-range applications
Ku	12-18 GHz	2.4-1.7 cm	Satellite altimetry
X	8-12 GHz	3.8-2.4 cm	Urban monitoring, snow cover
C	4-8 GHz	7.5-3.8 cm	Mapping, change detection
S	2-4 GHz	15-7.5 cm	Agriculture monitoring
L	1-2 GHz	30-15 cm	Geophysical and vegetation monitoring, InSAR
P	0.3-1 GHz	100-30 cm	Biomass monitoring

[1]https://www.earthdata.nasa.gov/learn/backgrounders/what-is-sar

8.3.2 DISTANCE OF THE TARGET FROM THE SATELLITE

When using RADAR systems to measure the distance to a target, the fundamental quantity is the time delay of the radio wave's round trip. If a target is located at a distance d from the RADAR system, and the radio wave travels to the target and back in t seconds, the total distance traveled by the beam is $2d$. The speed of the radio wave, c, is essentially the speed of light, since minor reductions in speed due to atmospheric effects are generally negligible for most practical applications. Therefore, the relationship between these variables can be expressed as:

$$c \times t = 2 \times d$$

Solving for d gives:

$$d = \frac{c \times t}{2}$$

This equation allows us to accurately compute the distance between the RADAR and the target. Beyond simple distance measurement, RADAR systems can also exploit other characteristics of the reflected signal, such as changes in polarization, to gain more detailed information about the target. This is where advanced RADAR techniques like Polarimetric SAR (PolSAR) and Interferometric SAR (InSAR) come into play.

8.3.3 POLSAR

PolSAR, or Polarimetric Synthetic Aperture RADAR, is an advanced RADAR technique that leverages the polarization properties of electromagnetic waves. Electromagnetic waves are transverse waves, meaning that their electric field oscillates perpendicular to the direction of propagation. When these waves are produced such that all oscillations are confined to a specific plane, the waves are said to be polarized. The orientation of polarization can be manipulated by a RADAR system's signal generator to create oriented polarized waves.

When a polarized wave interacts with a target, the polarization state of the reflected wave may remain the same, or may change. The occurrence and amount of change can provide significant insights into the target's properties. PolSAR systems can generate waves and capture returning waves in various polarization configurations, typically represented as two alphabets:

1. **VV (Vertical-Vertical)**: The incident wave is vertically polarized, and the vertical component of the reflected wave is recorded.
2. **VH (Vertical-Horizontal)**: The incident wave is vertically polarized, and the horizontal component of the reflected wave is recorded.
3. **HV (Horizontal-Vertical)**: The incident wave is horizontally polarized, and the vertical component of the reflected wave is recorded.

4. **HH (Horizontal-Horizontal)**: The incident wave is horizontally polarized, and the horizontal component of the reflected wave is recorded.

These configurations can provide a comprehensive view of the target's scattering properties, of which there are three main types [Figure 8.1a]:

1. **Rough-surface scattering**: This occurs when the surface texture is rough relative to the wavelength of the incident beam. A surface is considered rough if it meets the Fraunhofer criterion:

$$sd_h > \frac{\lambda}{32 \times \cos(\theta_i)}$$

   where sd_h represents the standard deviation of the surface height from its mean, λ is the wavelength of the incident beam, and θ_i is the incident angle. For rough-surface scattering, the relative scattering strengths are typically:

$$|S_{VV}| > |S_{HH}| > |S_{VH}| \text{ or } |S_{HV}|$$

   Examples of rough scatterers include agricultural fields with low vegetation, bare soil, and paved surfaces such as roads. As a surface becomes smoother relative to the wavelength, it may transition from rough-surface scattering to a smooth specular reflector (like reflection from a mirror), in which case minimal or no scattering is detected by the RADAR (since little or no radio wave is received back).

2. **Double-bounce scattering**: This type of scattering is associated with tall and vertical structures, such as buildings, tall trees, and utility poles. The incident beam undergoes two reflections before returning to the sensor. For double-bounce scattering, the relative scattering strengths are generally:

$$|S_{HH}| > |S_{VV}| > |S_{VH}| \text{ or } |S_{HV}|$$

   This type of scattering is particularly useful for identifying vertical structures and understanding their spatial orientation.

3. **Volume scattering**: This type of scattering occurs in areas with complex structures such as vegetation canopies. The incident beam scatters multiple times within the volume of the target before returning to the detector. In volume scattering, a significant amount of the reflected signal becomes cross-polarized, making $|S_{HV}|$ and $|S_{VH}|$ dominant. This scattering mode is particularly useful for distinguishing between different types of vegetation and monitoring changes in forested areas [Figure 8.1b–f].

PolSAR's ability to analyze and interpret changes in polarization offers a powerful tool for remote sensing. By examining the reflected waves' polarization characteristics, PolSAR can provide valuable insights into the physical properties and conditions of various surface features, enhancing our capability to monitor and analyze environmental and man-made structures.

8.4 LIDAR

LiDAR, which stands for "Light Detection and Ranging," represents a sophisticated active remote sensing technology that utilizes LASER light to measure distances with exceptional accuracy. The term itself reflects the integration of LASER and RADAR principles, highlighting its ability to detect and measure distances using light. Developed by Hughes Aircraft, Inc. in the 1960s, LiDAR has evolved into a critical tool in remote sensing, offering capabilities far beyond traditional optical and RADAR methods.

8.4.1 FEATURES OF LIDAR

LiDAR utilizes monochromatic light. Lasers emit light at a singular, precise wavelength, conferring LiDAR with exceptionally high spectral resolution. This attribute enables the system to discern subtle variations in surface materials based on their interaction with the LASER light. In topographic LiDAR applications, near-infrared lasers are employed to penetrate vegetation and capture intricate land surface features. In bathymetric LiDAR, green lasers are utilized to measure underwater features, such as seafloor and riverbed elevations, owing to their ability to penetrate water.

The LASER beam within a LiDAR system possesses remarkable directionality, maintaining its strength and focus over extended distances. This characteristic facilitates the detection and measurement of reflected signals from considerable distances, ensuring high-quality data acquisition. The narrow beam width also contributes to the reduction of measurement errors and enhances spatial resolution, rendering LiDAR highly effective for comprehensive and detailed surveys.

LiDAR systems exhibit exceptional precision and resolution, which are paramount in applications necessitating precise measurements. This precision is achieved through the meticulous measurement of time intervals between the emission and reception of LASER pulses, in conjunction with precise positional data obtained from Global Positioning System (GPS) and inertial measurement systems. The culmination of these efforts yields highly accurate spatial data, which can be utilized for a diverse range of applications, including topographic mapping and vegetation analysis.

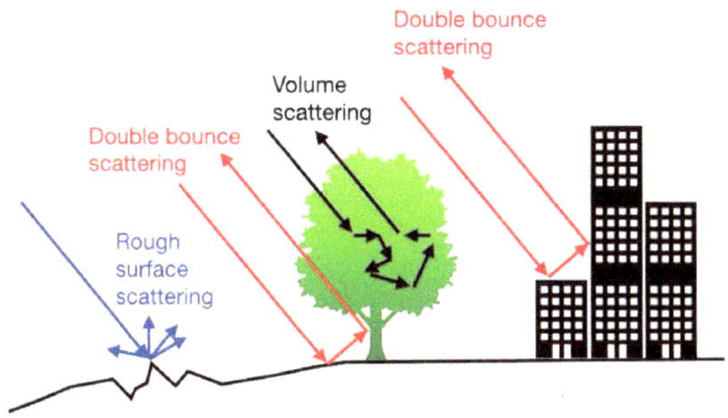

(a) Three kinds of scatterings.

(b) SAR deforestation alerts in Meghalaya due to slash-and-burn shifting cultivation (Jhum cultivation). The channels are R: 2023, G: 2022, B: 2021.

(c) A zoomed-up patch from (b), using same channels.

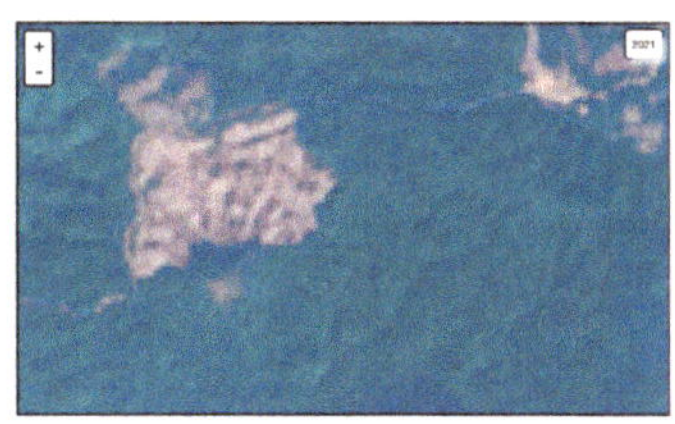

(d) Cloud-free optical imagery from year 2021 of the same area.

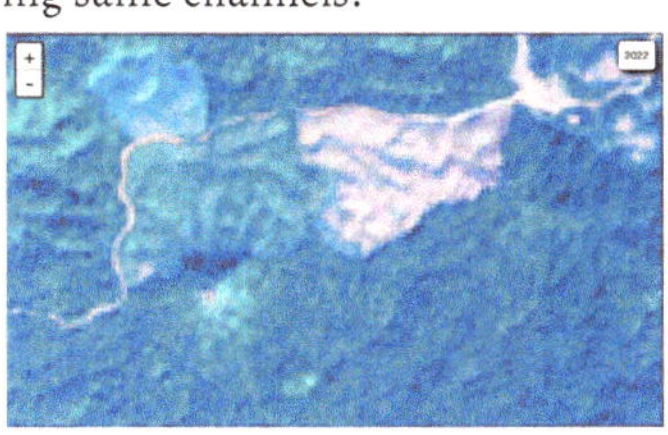

(e) Cloud-free optical imagery from year 2022 of the same area.

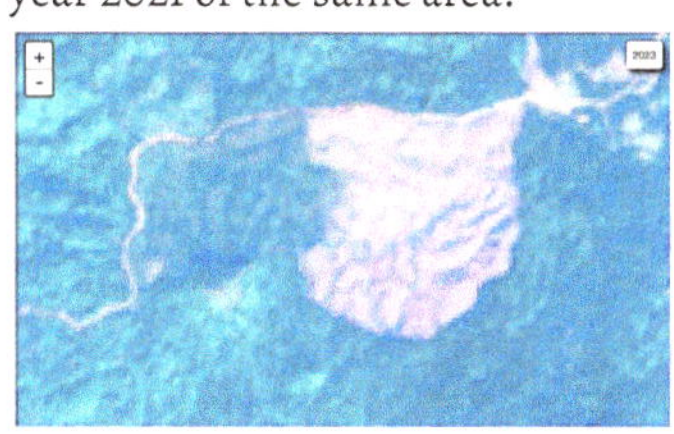

(f) Cloud-free optical imagery from year 2023 of the same area.

FIGURE 8.1: Use of SAR to detect forests and changes in forests. Data courtesy ESA and OSM. Visualized on geemap using Google Earth Engine.

8.4.2 FUNDAMENTAL COMPONENTS

To get information about a target, we require three measurements — the position of the LiDAR instrument and the distance and angular measurements of the target from the LiDAR instrument's position. The exact location of the LiDAR platform — whether mounted on an aircraft, UAV, or a ground-based system — is determined using Differential GPS (DGPS) and an Inertial Measurement Unit (IMU). DGPS provides accurate geographic coordinates, while the IMU measures orientation and movement. Together, these systems ensure that the LiDAR data is accurately georeferenced, allowing for precise spatial analysis.

The distance from the LiDAR instrument to the target is computed using the same formula that we discussed in RADAR distance measurement:

$$d = \frac{c \times t}{2}$$

In this formula, d is the distance to the target, c represents the speed of light ($299,792,458$ meters per second in a vacuum, with a slight reduction in refractive media like air), and t is the time it takes for the LASER pulse to travel to the target and return. The factor of 2 accounts for the round-trip distance traveled by the LASER beam.

The angle at which the LASER beam is emitted and received is also crucial for creating three-dimensional (3D) models of the target area. By measuring these angles, LiDAR systems can generate detailed 3D representations of landscapes and structures, capturing intricate details that are essential for accurate spatial analysis.

8.4.3 MODES OF OPERATION

LiDAR systems can be operated in various modes to capture different types of data:

1. **First Pulse (FP) Mode**: In FP mode, the LiDAR system detects the initial pulse of the reflected LASER beam. This pulse typically corresponds to the elevation of the highest surfaces within the target area, such as tree canopies or building rooftops. FP mode is particularly advantageous for capturing the uppermost features and provides valuable information for applications such as vegetation mapping and urban planning.

2. **Last Pulse (LP) Mode**: LP mode detects the last pulse of the reflected LASER beam, which typically corresponds to the ground surface elevation. This mode provides a comprehensive view of the terrain, enabling the generation of essential data for topographic mapping and ground surface analysis. LP mode effectively mitigates the influence of overlying vegetation or structures, thereby offering a more accurate representation of the underlying terrain.

These modes can be employed individually or in conjunction to generate distinct data models:

1. **Digital Elevation Model (DEM)**: The DEM represents the elevation of the tallest surfaces detected by the first pulse. DEMs are valuable for applications such as flood modeling, land use planning, and environmental assessment.
2. **Digital Terrain Model (DTM)**: The DTM depicts the ground surface elevation, derived from the last pulse mode. It provides an accurate representation of the terrain without interference from vegetation or structures. DTMs are used for applications such as engineering design, infrastructure development, and geospatial analysis.
3. **Digital Canopy Height Model (DCHM)**: The DCHM is calculated by subtracting the DTM from the DEM, representing the height of vegetation or other objects above the ground. This model is useful for analyzing forest canopy height, assessing vegetation density, and studying habitat structure. DCHMs provide insights into forest health, biomass, carbon sequestration, and ecological dynamics.

8.4.4 APPLICATIONS IN FORESTRY

LiDAR technology has proven to be exceptionally versatile and valuable in forestry and ecological studies, offering numerous applications.

1. **Aerial and space-based surveys**: LiDAR systems are deployed on various platforms, including aircraft, UAVs, and space-based missions such as NASA's GEDI (Global Ecosystem Dynamics Investigation) aboard the International Space Station [Figure 8.2a]. This deployment option allows for data collection at both local and global scales, enabling comprehensive forest monitoring and analysis.
2. **Topographic mapping**: By utilizing FP and LP modes, LiDAR generates highly accurate DEMs and DTMs, providing detailed information on ground topography and forest canopy heights [Figure 8.2b, c]. This capability is crucial for understanding the structure and dynamics of forest ecosystems, assessing landform changes, and supporting sustainable land management practices.
3. **Biomass estimation**: LiDAR data can be used to estimate biomass density in forest stands by analyzing canopy-penetrating LASER beams [Figure 8.2d–g]. This information supports the creation of detailed canopy structure models, offering insights into forest carbon storage, productivity, and ecological health. Accurate biomass estimates are essential for carbon accounting and forest conservation efforts.

4. **Monitoring changes in forests**: LiDAR's ability to provide continuous, high-resolution measurements enables the monitoring of changes in forests over time. This capability is valuable for tracking the effects of disturbances such as wildfires, logging, and disease outbreaks. LiDAR data helps researchers and policymakers assess forest recovery, evaluate the impacts of environmental changes, and plan effective conservation strategies.
5. **Terrestrial Laser Scanning (TLS)**: LiDAR systems can also be used in a horizontal manner as Terrestrial Laser Scanners (TLS), which are either mounted on tripods or UAVs for deployment in dense forest areas. TLS provides detailed, high-resolution information about forest attributes at a more localized scale, enhancing our understanding of forest structure, species composition, and habitat characteristics [Figure 8.2h].

In this way, LiDAR technology has revolutionized the field of remote sensing by providing precise, high-resolution spatial data. Its applications span a wide range of fields, from forestry and environmental monitoring to urban planning and infrastructure development. The ability to deliver detailed 3D models and accurate measurements makes LiDAR an indispensable tool for advancing our understanding of natural and built environments, supporting informed decision-making, and fostering sustainable resource management.

8.5 GEOGRAPHIC INFORMATION SYSTEMS (GIS)

Geographic Information Systems (GIS) are sophisticated frameworks that integrate hardware, software, and human expertise to capture, store, manipulate, analyze, manage, and present spatial data. Unlike traditional methods of analyzing geographic data, which involved physical maps and manual overlays, modern GIS employs digital technology to create layered digital representations of spatial information. This digital approach allows for more precise analysis and visualization of geographic phenomena.

In the contemporary era of technological advancement, they have become indispensable tools for managing and analyzing spatial data. The rise of GIS technology has revolutionized how we interact with and understand geographic information, providing profound insights across a multitude of fields including urban planning, environmental management, and disaster response.

8.5.1 CORE COMPONENTS OF GIS

1. **Hardware** comprises:
   - **Computers**: High-performance systems designed to handle extensive datasets and complex analytical tasks.
   - **GPS devices**: Instruments used to acquire accurate geographic coordinates.

(a) LiDAR can be used even at global scales.

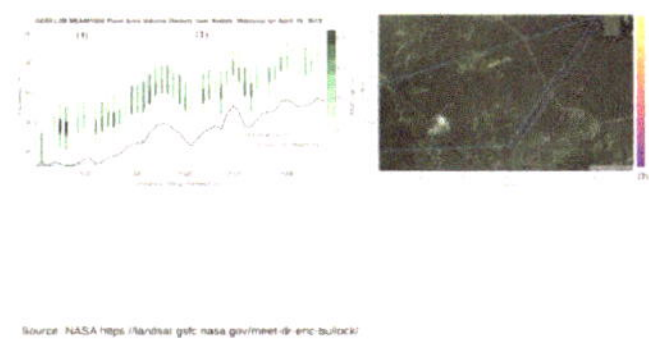

(b) FP and LP are used to compute topography and canopy height.

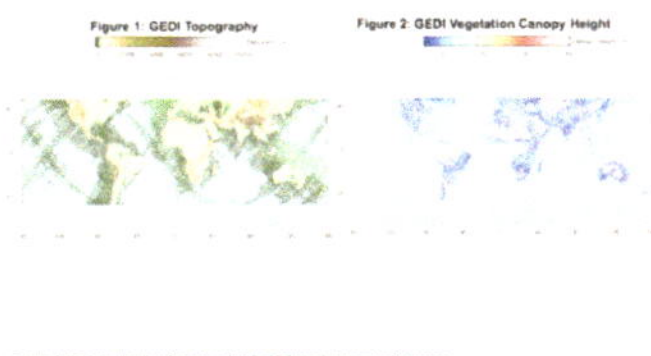

(c) Global-scale topography and canopy measurements can be done.

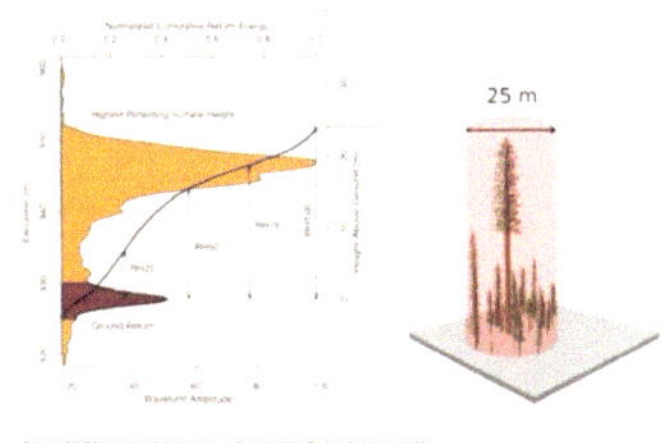

(d) Biomass density can be computed using returning pulse information.

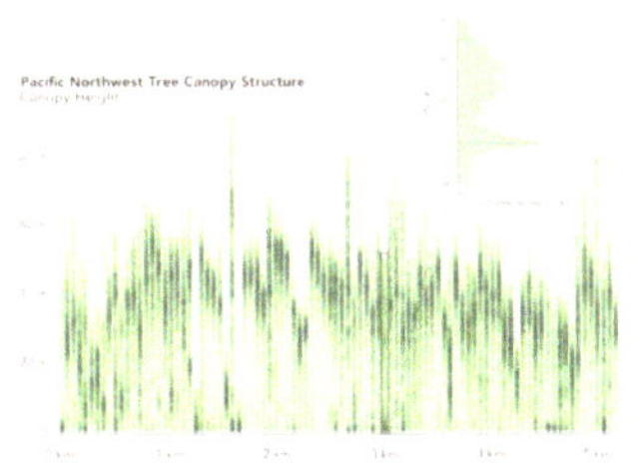

(e) Elaborate canopy structure models can be developed.

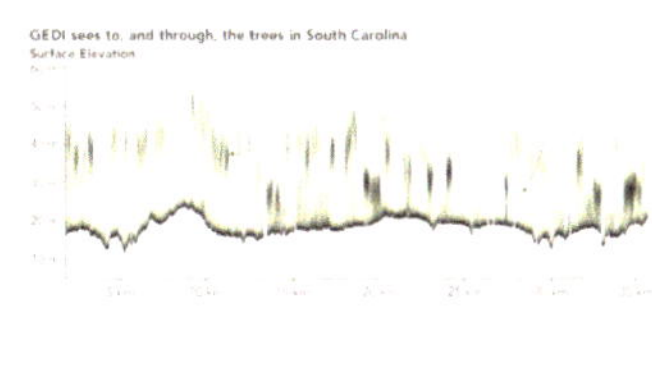

(f) This permits looking through the trees in the forest.

(g) Hence very detailed biomass maps can be generated.

(h) LiDAR can also be used horizontally for specific applications.

FIGURE 8.2: Some uses of LiDAR in forestry.

   - **Sensors**: Devices that collect environmental data, such as satellite and ground-based sensors, crucial for gathering spatial information.
   - **Servers**: Systems that store and manage GIS data, facilitating processing and access.
2. **Software** includes:
   - **QGIS**: An open-source GIS software known for its comprehensive functionality and user-friendly interface. QGIS supports various spatial analysis tasks and is highly extensible through plugins.
   - **GRASS GIS**: Another open-source GIS tool, valued for its advanced geospatial analysis capabilities and data modeling support.
   - **ArcGIS**: A commercial GIS software suite developed by Esri, providing a broad range of tools for spatial analysis, mapping, and data management. It is widely utilized in various industries for its extensive features and robust support.
   - **Erdas Imagine**: A commercial GIS software suite developed by Hexagon Geospatial, used for processing and analyzing geospatial data such as satellite imagery. It offers tools for image classification, terrain modeling, and spatial analysis, widely used in fields like environmental monitoring and urban planning.
3. **Humanware** encompasses:
   - **GIS professionals**: Specialists who design, implement, and manage GIS projects, ensuring effective data integration and analysis.
   - **Data analysts**: Individuals who interpret GIS data to derive meaningful insights and support decision-making processes.
   - **Decision-makers**: Stakeholders who utilize GIS outputs to make informed decisions in fields such as urban planning, environmental conservation, and disaster management.

8.5.2 DATA INTEGRATION IN GIS

Modern GIS operates through a layered digital framework where geographic data is represented in discrete layers. Each layer may represent different types of data, such as land use, elevation, or infrastructure. By integrating and analyzing these layers, GIS enables the exploration of complex spatial relationships and patterns. For instance, combining land use data with demographic information can provide insights into urban growth and infrastructure needs.

A key strength of GIS is its ability to integrate diverse data sources, enhancing spatial analysis. Sources of data include:

1. **Satellite imagery**: Provides extensive coverage and detailed information on land cover and land use. GIS integrates satellite images with ground-truth data from field surveys to create accurate and comprehensive maps.

2. **Aerial photographs**: Offer high-resolution views of specific areas, useful for updating and refining GIS data layers.

3. **Field surveys**: Provide on-the-ground data to validate and enhance remote sensing information, including vegetation assessments, soil sampling, and infrastructure inspections.
4. **Scanned maps and toposheets**: Historical maps and topographic sheets offer valuable context and historical data.
5. **GPS coordinates**: Precise location data collected through Global Positioning System technology.
6. **Photographs and documents**: Historical photographs and documents provide additional contextual information.

8.5.3 APPLICATIONS IN FORESTRY AND WILDFIRE MANAGEMENT

Geographic Information Systems play a pivotal role in both forestry management and wildfire detection and response. In forestry management, GIS facilitates the creation of comprehensive maps of forested areas, encompassing tree species, age classes, and health conditions. These maps serve as indispensable tools for effective forest management and conservation planning. Furthermore, the integration of LiDAR data with GIS provides precise estimates of forest biomass, which is crucial for carbon accounting, ecological assessment, and resource management. GIS also supports the analysis and mapping of wildlife habitats within forests, facilitating conservation efforts by identifying critical habitats and monitoring changes over time.

In the realm of wildfire detection and management, GIS assists in analyzing factors that contribute to wildfire risk, such as vegetation types, slope, and historical fire patterns. This analysis enables the identification of high-risk areas and the development of targeted mitigation strategies. During a wildfire, GIS integrates data from satellite sensors, weather stations, and ground reports to provide real-time updates on fire behavior, spread, and containment efforts. This information is pivotal for coordinating emergency responses and resource allocation. Post-wildfire, GIS is utilized to assess damage, evaluate the effectiveness of fire management strategies, and plan recovery and rehabilitation efforts. It assists in mapping affected areas, tracking recovery progress, and supporting long-term ecological monitoring.

8.6 DRONES

Drones, also known as Unmanned Aerial Vehicles (UAVs), are aircraft that operate without a human pilot onboard. These autonomous or remotely controlled devices represent a substantial advancement in remote sensing technology, providing a novel platform for capturing aerial data. In contrast to conventional methods that entail the utilization of large aircraft, balloons, or satellites, drones possess the capability to operate at considerably lower altitudes, thereby enabling the acquisition of imagery with enhanced spatial resolution.

8.6.1 WORKING OF DRONES

Drones function through a harmonious integration of hardware and software components, meticulously designed to efficiently capture and transmit data. The foundational elements of this system encompass the airframe, flight control system, navigation and communication protocols, sensors and cameras, and power supply.

The airframe is the structural framework of the drone, typically made from lightweight yet durable materials such as carbon fiber or plastic. It supports the drone's propulsion system, which consists of rotors or propellers that generate lift and enable flight. Fixed-wing drones utilize aerodynamic surfaces to maintain flight over longer distances.

The flight control system acts as the drone's brain, processing input from various sensors and executing control commands to ensure stability and precise navigation. It includes gyroscopes and accelerometers that measure orientation and movement to maintain smooth and controlled flight.

Navigation and communication are critical for drone operation. Global Positioning System (GPS) technology provides location data, allowing the drone to follow predefined flight paths and maintain altitude. The communication system utilizes radio frequency (RF) to facilitate data exchange between the drone and the operator, encompassing control commands and telemetry information. Additionally, certain drones provide live video streaming capabilities, which are crucial for real-time monitoring and visualization.

Drones can be equipped with a variety of sensors and cameras, including high-resolution optical cameras, LiDAR sensors, and thermal imaging cameras.

The primary power source for drones is usually rechargeable batteries that provide power to the motors, sensors, and communication systems. Battery life is a crucial factor in determining the drone's operational range and flight duration.

Software plays a crucial role in drone operations. Flight planning and control software allows operators to program flight paths, set waypoints, and manage the drone's movements [Figure 8.3b]. Advanced software may offer autonomous flight capabilities and real-time data processing. Data processing software is used to analyze and interpret the data collected by the drone's sensors, including stitching images together and generating maps or models.

8.6.2 ADVANTAGES OF DRONES

Drones offer several advantages over conventional aerial data collection techniques. Their ability to operate at low altitudes allows for the acquisition of high-resolution imagery, revealing finer details than satellite imagery. This enhanced spatial resolution is particularly beneficial for comprehensive forest inventory and mapping.

Drones provide flexibility in flight altitude, which directly influences spatial resolution and areal coverage. Flying at higher altitudes results in lower spatial resolution

(a) Satellite imagery of a location for drone deployment.

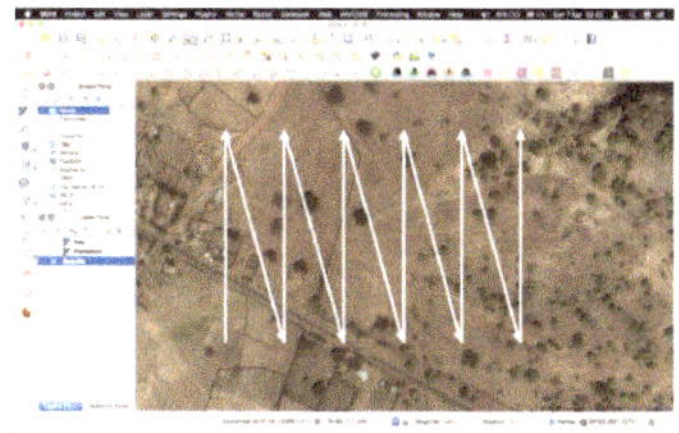

(b) Flight plans can be programmed into many drones.

(c) At greater heights, more area can be covered with less spatial resolution.

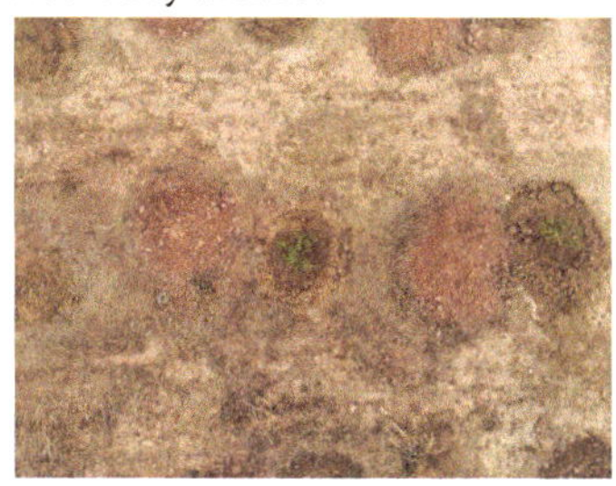

(d) At lower heights, less area is covered with high spatial resolution.

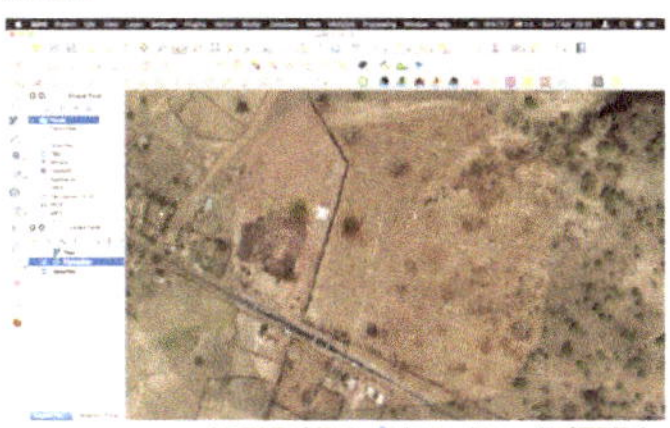

(e) Overlapping drone images can be stitched into a single large georeferenced image.

(f) This permits the discernment of changes that have occurred in the elapsed time.

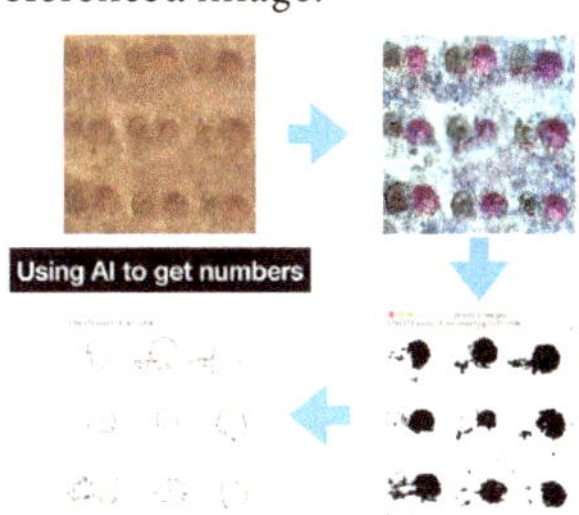

(g) AI algorithm can be used to automatically analyze images.

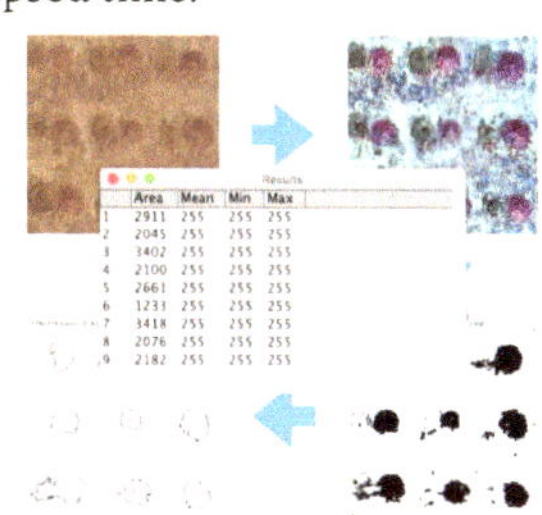

(h) The size of each individual dug pit can be computed automatically.

FIGURE 8.3: Some uses of drones in forest remote sensing.

but covers a larger area [Figure 8.3c], while lower altitudes offer higher resolution but cover a smaller area [Figure 8.3d]. This flexibility allows for targeted data collection based on specific needs. At very low heights, drones can capture detailed imagery that is essential for applications such as distinguishing individual leaves and assessing fine-scale vegetation changes [Figure 8.3d].

One of the key advantages of drones is their ability to capture and stitch together multiple images into a seamless, geo-referenced map [Figure 8.3e]. This process involves programming sufficient overlap between adjacent images, which is then used by computer software to create a continuous map. The stitched image not only confirms that the images are from the same location but also allows for the assessment of changes over time, such as detecting alterations in forest cover or infrastructure [Figure 8.3f].

Automated analysis of stitched images can further enhance the utility of drone data. For example, contrast-enhanced images can be processed by artificial intelligence algorithms to identify and quantify specific features, such as planting pits [Figure 8.3g–h]. These techniques can also be applied to wildfire management, where drones help delineate affected areas and evaluate the effectiveness of response measures.

8.6.3 APPLICATIONS OF DRONES

In the field of forestry, drones serve as indispensable tools for a wide range of applications. They offer high-resolution imagery, facilitating the meticulous mapping of forest areas. This imagery facilitates the inventory of tree species, their age, and distribution. Furthermore, the utilization of LiDAR sensors enables the generation of three-dimensional data on forest structure, encompassing canopy height and tree density. This data is crucial for evaluating forest biomass and volume.

Drones equipped with multispectral sensors are frequently utilized for monitoring vegetation health. These sensors analyze reflectance in various spectral bands, enabling the identification of stress, diseases, or pest infestations in trees. Indices such as the Normalized Difference Vegetation Index (NDVI), derived from drone imagery, are used to assess vegetation vigor and productivity, supporting effective forest management and conservation efforts.

In the realm of wildlife monitoring, drones provide a non-invasive method for tracking animal movements, nesting sites, and population densities. They facilitate habitat assessment by mapping critical areas and monitoring changes over time. In wildlife management, drones are employed for animal herding and darting operations.

Drones also play a pivotal role in forest health assessment. They facilitate rapid damage assessments following storms or fires, aiding in recovery planning and evaluating the efficacy of response strategies. Regular drone flights enable the timely detection of early indications of diseases or pests, facilitating prompt intervention.

Drones are particularly useful in wildfire management, offering capabilities that enhance pre-fire risk assessments, real-time detection and monitoring, and post-fire recovery. Prior to a wildfire, drones can conduct risk assessments by mapping areas vulnerable to fire and identifying potential hazards, aiding in wildfire prevention efforts.

During a wildfire, drones equipped with thermal and optical sensors can provide real-time data on fire spread, intensity, and behavior. Live video streaming from drones enhances situational awareness and coordination among firefighting teams, while thermal imaging helps detect hotspots and smoldering areas, guiding firefighting efforts.

After a wildfire, drones are used to map burned areas and assess damage. This information is paramount for recovery planning and assessing the efficacy of fire management strategies. Drones also support post-fire rehabilitation by evaluating soil erosion, monitoring vegetation recovery, and ensuring that recovery efforts are successful.

Furthermore, drones contribute to public safety by providing aerial perspectives of hazardous areas, safeguarding first responders, and disseminating information regarding fire safety and recovery initiatives to the public.

In this manner, drones have revolutionized the domains of forestry and wildfire management by offering high-resolution imagery, real-time monitoring capabilities, and sophisticated data collection methodologies. Their capacity to operate at low altitudes and capture detailed data enhances our comprehension and management of forest resources and wildfire incidents. As drone technology advances, its applications are anticipated to expand, providing even greater opportunities for enhancing forest management and wildfire response.

9 Computational systems in wildfire management

Wildfires pose substantial challenges to emergency response and management, necessitating the adoption of sophisticated technological solutions to predict, monitor, and mitigate their impacts. For these, robust computational systems, including fire behavior modeling, decision support systems (DSS), and emerging technologies such as the Internet of Things (IoT), artificial intelligence (AI), machine learning (ML), and cloud computing, become necessary. This chapter delves into the intricacies of these systems and elucidates their pivotal role in refining wildfire management strategies.

9.1 FIRE BEHAVIOR MODELING

Fire behavior modeling is an essential component of wildfire management, providing the necessary insights to understand and predict how fires spread under various conditions (Or et al., 2023). This modeling incorporates advanced algorithms and simulations that integrate various factors, including weather conditions, terrain, and fuel types. Through the utilization of these models, fire managers can forecast fire behavior, devise effective strategies, and allocate resources in a coordinated manner.

Fire spread models are instrumental in predicting how wildfires advance across different landscapes. They can be categorized into physical and empirical models, each offering distinct advantages in understanding fire dynamics.

Physical models are grounded in the fundamental principles of fire physics. These models employ mathematical equations to simulate the processes of heat transfer, combustion, and fire dynamics. Heat transfer mechanisms, encompassing conduction, convection, and radiation, are fundamental to physical models. By integrating these mechanisms, physical models provide a comprehensive understanding of how heat interacts with fuels and the surrounding environment. Furthermore, these models incorporate the chemical reactions associated with combustion, such as the breakdown of fuel molecules and the generation of combustion gases. These reactions influence the combustion behavior of various fuels and the distribution of heat. Physical models serve as valuable tools for scientific research and advanced analysis, offering insights into fire behavior under specific conditions and contributing to a deeper comprehension of the factors that drive fire spread.

Empirical models, in contrast, rely on historical data and observed fire behavior to make predictions. These models employ statistical techniques to analyze past fire incidents and discern patterns and relationships. By examining historical data, such as fire size, spread rates, and fuel conditions, empirical models establish statistical connections that can be applied to forecast future fire behavior. These models are particularly valuable for operational decision-making, offering rapid, data-driven assessments of fire behavior based on past experiences. They are also beneficial for assessing fire risk by identifying areas with a higher probability of fire occurrence. Empirical models are frequently validated against historical fire data to ensure accuracy and are subsequently refined through feedback from actual fire events.

Among the various models used, FARSITE (Fire Area Simulator) and BEHAVE are two prominent operational tools that offer valuable predictions of fire behavior.

FARSITE (Fire Area Simulator) (Finney, 1998) is a comprehensive fire behavior simulation tool developed by the U.S. Forest Service. The FARSITE model incorporates a comprehensive range of variables to simulate the spatial and temporal dynamics of fire spread. These variables include detailed meteorological data such as wind speed and direction, temperature, and humidity, which significantly influence fire behavior. For instance, winds can propel flames and embers to new locations, while temperature and humidity affect fuel moisture content and combustion rates. Additionally, the model considers topographical features such as slope and aspect. Steeper slopes can accelerate fire spread due to the preheating of fuels above the fire, and the aspect of slopes influences solar exposure and vegetation growth. The model also requires data on fuel types, including vegetation density and composition, as different fuels burn at varying rates and intensities. FARSITE's strength lies in its ability to simulate intricate interactions over time, providing detailed predictions of fire perimeters, intensity, and potential impacts. This makes it an indispensable tool for operational planning and strategic decision-making in wildfire management.

BEHAVE (Burgan, 1984), another critical fire behavior prediction system, focuses on user-defined parameters, allowing for rapid assessments based on immediate observations. BEHAVE incorporates inputs related to fuel types, slope, and wind conditions. By entering specific data about various fuel types, such as grass, shrubs, or timber, users can estimate key fire behavior metrics, including rate of spread, flame length, and fireline intensity. These metrics are crucial for comprehending the potential impact of a fire and for making informed decisions regarding fire suppression and resource allocation. BEHAVE is particularly valuable in the initial stages of fire management, offering rapid and actionable insights that facilitate the formulation of initial attack strategies and resource planning.

9.2 COMPUTATIONAL SYSTEMS FOR REMOTE SENSING AND GIS DATA ANALYSES

The utilization of computational systems in conjunction with remote sensing and Geographic Information Systems (GIS) has fundamentally transformed wildfire management practices. These technologies, when integrated, offer comprehensive tools for monitoring, analyzing, and responding to wildfires. This section delves into the manner in which computational systems augment the capabilities of remote sensing and Geographic Information Systems (GIS). It provides a comprehensive overview of their applications and the benefits they confer on wildfire management.

9.2.1 REMOTE SENSING INTEGRATION

Remote sensing technologies capture critical data on wildfires from satellite and airborne platforms. Computational systems significantly enhance the utility of this data through advanced processing, analysis, and integration. For instance, image processing techniques such as classification and segmentation are employed to identify fire perimeters, detect hot spots, and assess burn severity. Computational systems can efficiently handle this processing at scale and speed, ensuring that information is readily available in near real-time.

Computational systems are also instrumental in analyzing temporal changes by comparing sequential remote sensing images. This change detection capability is paramount for tracking fire progression and evaluating the efficacy of firefighting efforts. Algorithms that identify alterations in vegetation health, fire perimeter expansion, and thermal anomalies provide a dynamic perspective of the fire's behavior, enabling responsive and adaptive management strategies.

The ability to process and integrate real-time remote sensing data into computational systems is paramount for timely decision-making. These systems can continuously update fire models and maps with the most recent data, providing current information on fire location, intensity, and spread. Real-time integration enables fire managers to make informed decisions swiftly, optimizing response efforts and resource allocation. We have observed an example of such integration in the form of the SimplyFire[1] application, which provides continuously updated firc data in a graphical format.

9.2.2 GIS DATA ANALYSES

GIS technologies, when enhanced by computational systems, offer powerful tools for spatial data analysis and visualization. This integration supports various aspects of wildfire management:

Spatial Data Integration and Visualization: Geographic Information Systems (GIS) platforms, supported by computational systems, integrate various data layers, including topography, fuel types, weather conditions, and infrastructure. This integration facilitates the creation of detailed maps that depict the spatial distribution of wildfires. Computational systems process and analyze these data layers to generate comprehensive visualizations that facilitate the comprehension of fire behavior

[1]http://mpforest.gov.in/publicdomain/atlas/index.html

and impacts. Dynamic mapping tools enable the creation of interactive maps that can be updated with new data, providing a near real-time perspective of the fire situation.

Advanced Fire Behavior Modeling: Computational systems augment GIS-based fire behavior models by facilitating intricate simulations of fire propagation and intensity. These models incorporate diverse inputs, such as fuel characteristics, meteorological conditions, and topographical features, to forecast the evolution of fires over time. Computational systems execute these simulations with remarkable efficiency, enabling the exploration of various scenarios and assisting fire managers in anticipating potential fire behavior and impacts.

Risk assessment and planning: GIS, powered by computational systems, is used to assess fire risk and develop strategic plans (Awadhiya, 2018). Risk assessment models analyze data on fuel loads, weather patterns, and historical fire behavior to identify high-risk areas. Computational systems support these models by processing vast datasets and executing simulations to forecast fire risk and guide mitigation strategies. This proactive approach enables the prioritization of areas for fuel reduction, the development of evacuation plans, and the efficient allocation of resources.

9.3 PREDICTIVE ANALYTICS

Predictive analytics, bolstered by advancements in artificial intelligence (AI) and machine learning (ML), has become a cornerstone in wildfire management. This approach leverages historical data, real-time inputs, and sophisticated algorithms to enhance fire spread modeling, risk assessment, and strategic decision-making. By harnessing these technologies, fire managers can forecast fire behavior, identify risks, and implement proactive measures to mitigate wildfire impacts.

9.3.1 FORECASTING

Predictive analytics tools harness historical data, real-time observations, and intricate algorithms to forecast wildfire behavior. This forecasting capability is paramount for comprehending the potential spread of fires and for devising effective management strategies in various ways.

Historical data analysis in predictive modeling: Predictive models are heavily reliant on extensive historical data, encompassing past fire occurrences, weather conditions, and fuel types. Through the analysis of historical fire behavior, these models can discern patterns and trends that serve as the foundation for future predictions. Historical data can shed light on how specific weather conditions or fuel types have historically impacted fire spread, thereby providing invaluable insights for current forecasting endeavors.

One such example is presented in figure 9.1 that explores the influence of agricultural waste burning on wildfire initiation. It is common knowledge that winds, especially local winds travelling from agricultural fields towards forest areas, can transport burning particles generated when agricultural wastes are burnt [Figure 9.1a]. These burning particles can easily initiate new wildfires, especially in dry conditions or when there is an abundance of dried leaf litter on the forest floor [Figure 9.1b].

This model can easily be corroborated by constructing burn severity maps [Figure 9.1c-III] using pre-fire [Figure 9.1c-I] and post-fire [Figure 9.1c-II] satellite data. Regions with sharp boundaries between agricultural and forest areas [Figure 9.1c-IV] often translate into corresponding sharp fire signatures [Figure 9.1c-V], and in many cases, the spread of agricultural fires to nearby forest areas can also be witnessed [Figure 9.1c-VI].

When we have observed such correlations over several years through historical data analyses and have become certain of causality factors, we can incorporate our understanding into wildfire management programs, such as by providing additional resources near such agricultural-forest land interface locations.

Real-time data integration: Integrating real-time data, including current weather conditions, satellite imagery, and sensor readings, significantly enhances the accuracy of fire behavior forecasts. Predictive analytics systems process this data to dynamically update fire spread models, providing real-time predictions of fire progression. This integration enables fire managers to respond promptly to evolving conditions and adjust strategies accordingly.

One such integration is depicted in figure 9.2 where vegetation, slope, wind, and historical wildfire data are made use of for predictive forecasting on a daily basis by integrating day-to-day weather data into the model.

Algorithmic forecasting: Advanced algorithms, particularly those incorporating artificial intelligence (AI) and machine learning (ML), enhance forecasting accuracy by analyzing intricate interactions between various factors influencing fire behavior. Machine learning models can acquire knowledge from historical data and refine their predictions based on novel inputs. These algorithms simulate diverse scenarios, encompassing varying weather conditions and fuel types, to forecast potential fire outcomes and evaluate the impact of various management strategies.

9.3.2 RISK IDENTIFICATION

Machine learning algorithms excel at analyzing extensive and diverse datasets to discern patterns and anomalies that may suggest heightened wildfire risk. This capability facilitates early detection and targeted prevention measures through early warning systems in various ways.

Pattern recognition: Machine learning algorithms can discern patterns in data that indicate elevated fire risk. For instance, these algorithms can analyze weather patterns, vegetation conditions, and historical fire data to identify unusual trends or anomalies that suggest heightened wildfire potential. By recognizing these pat-

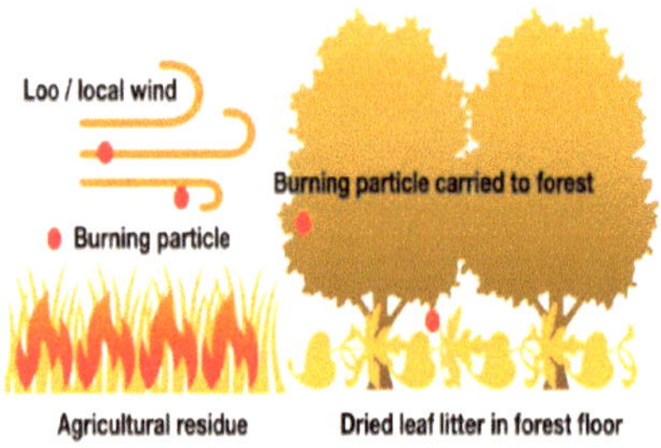

(a) Wind movement from agricultural fields towards forests has the ability to transport burning particles.

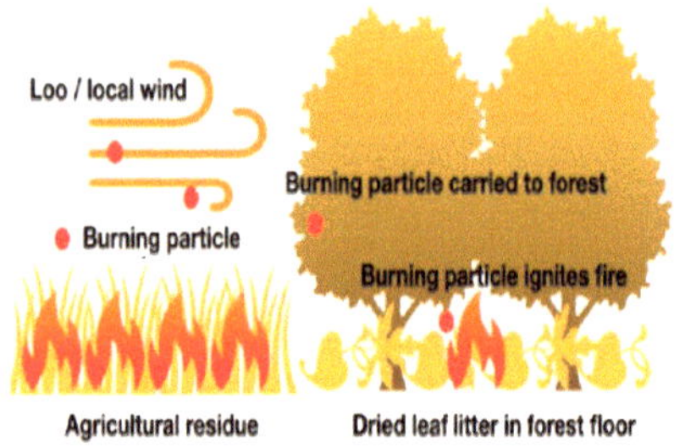

(b) Burning particles can initiate new wildfires, especially in areas with dried leaf litter and dry grasses.

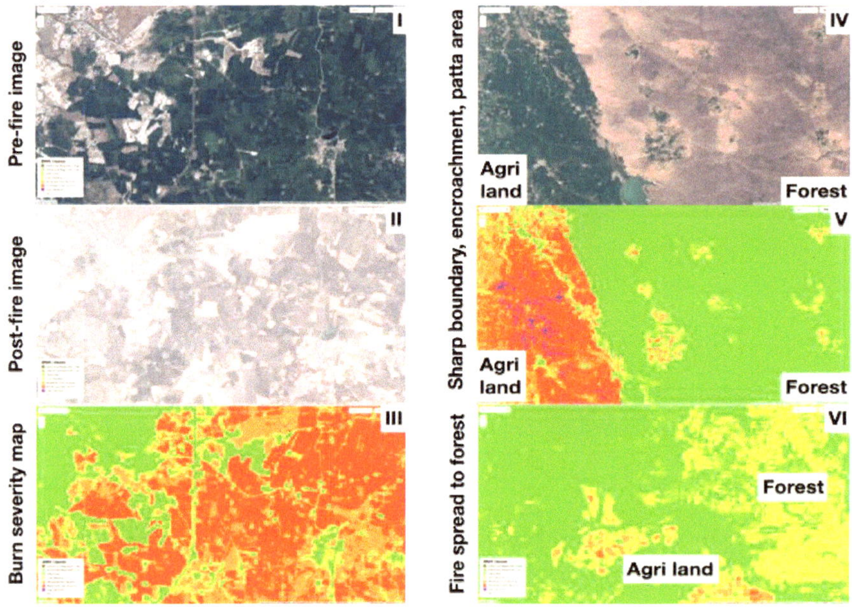

(c) Places with fires can easily be identified using satellite remote sensing and GIS. We can use pre-fire image (I) and post-fire image (II) to compute burn severity maps (III). In many areas, sharp boundaries between agricultural land and forests (IV) are reflected in burn severity maps (V) when agricultural wastes are burnt as a cultural practice. In many cases, we can trace wildfires to nearby agricultural lands from which they have spread (VI). Visualized on Google Earth Engine.

FIGURE 9.1: Influence of agricultural waste burning on wildfire initiation.

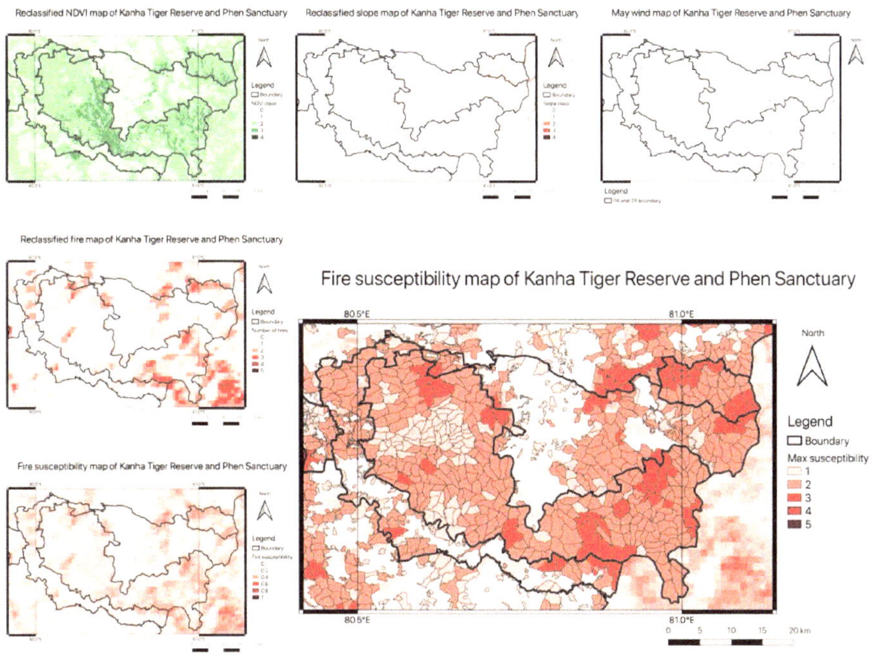

FIGURE 9.2: Predictive forecasting of wildfires using vegetation, slope, wind, and historical wildfire data.

terns, fire managers can prioritize high-risk areas and implement preventive measures.

Anomaly detection: Anomaly detection is a crucial technique employed in predictive analytics to identify outliers within data sets. These outliers may serve as indicators of potential emerging risks. For instance, abrupt fluctuations in vegetation moisture levels or anomalous temperature spikes can be recognized as anomalies that necessitate prompt investigation. Early identification of such anomalies facilitates timely intervention and mitigates the risk of wildfires escalating.

Early Warning Systems: Machine learning algorithms play a pivotal role in the development of early warning systems that provide advanced notice of potential wildfire outbreaks. Through continuous analysis of data from diverse sources, including remote sensing imagery, weather stations, and historical records, these systems can promptly issue alerts when environmental conditions are conducive to wildfire ignition. Early warnings enable prompt preventive measures, such as the implementation of burn bans or the preemptive deployment of firefighting resources.

9.4 DECISION SUPPORT SYSTEMS (DSS)

Decision Support Systems (DSS) are crucial tools in wildfire management, designed to integrate various data sources and support strategic decision-making (Martell, 2015). By integrating data from various sources, Decision Support Systems augment operational efficiency, optimize resource allocation, and facilitate the formulation of effective response strategies. This section delves into the salient components and applications of DSS in the management of wildfires.

Integrated data platforms serve as the foundation of Decision Support Systems, offering a comprehensive view of the wildfire situation by integrating information from diverse sources. DSS platforms aggregate data from fire behavior models, remote sensing technologies, Geographic Information Systems (GIS), and weather forecasts. By unifying these disparate data sources, DSS platforms provide a holistic perspective of the wildfire situation. This comprehensive view enables fire managers to assess current conditions, predict future developments, and make informed decisions based on the most up-to-date information. For instance, a DSS may integrate real-time satellite imagery with weather forecasts and GIS data to provide an updated map of the fire's progression, facilitating more effective planning and response.

This helps improve situational awareness by consolidating information into a single interface. This centralized perspective facilitates decision-makers in comprehending the interconnections between various factors influencing the fire, including the potential impact of alterations in wind direction on the spread of the fire across specific terrain.

9.4.1 SCENARIO ANALYSIS

Scenario analysis is a fundamental feature of Decision Support Systems, enabling users to simulate diverse fire scenarios based on various assumptions. DSS can model these scenarios by modifying variables such as weather conditions, fuel types, and topographical features. For instance, a DSS might simulate how a fire would behave under extreme wind conditions compared to moderate winds, or how different fuel types would influence the fire's intensity and spread.

Through these simulations, DSS facilitates the evaluation of potential outcomes and the preparation for various contingencies. This capability empowers fire managers to explore alternative "what-if" scenarios and develop strategies for a range of possible situations. For example, if a model predicts rapid fire spread under specific conditions, fire managers can formulate contingency plans for additional resources or evacuation measures, thereby supporting strategic planning by providing insights into the potential impact of different decisions.

9.4.2 RESOURCE MANAGEMENT

Effective resource management is a fundamental aspect of wildfire response, and Decision Support Systems play a pivotal role in optimizing the allocation of firefighting resources. By analyzing fire behavior predictions in conjunction with data on resource availability, DSS can identify the most effective deployment strategies. For instance, if a fire behavior model forecasts a rapidly spreading fire in a specific area, the DSS can recommend the deployment of additional firefighting personnel, equipment, and supplies to that location.

Through the utilization of DSS, fire managers can ensure that resources are efficiently utilized and directed to where they are most required. This optimization mitigates the risk of resource shortages in critical areas and enhances the overall effectiveness of firefighting efforts.

DSS also facilitates logistics management by tracking resource movements, coordinating supply chains, and ensuring that resources are distributed in accordance with priority needs. This approach contributes to maintaining a balanced and responsive approach to wildfire management.

9.4.3 EVACUATION PLANNING

Evacuation planning is a fundamental aspect of wildfire management, and DSS provide invaluable support in developing effective evacuation strategies.

Through fire spread simulation, we can assess the potential impact of a wildfire on various areas, identifying vulnerable regions and devising evacuation routes accordingly.

DSS can also evaluate the safety and accessibility of evacuation routes by considering factors such as road conditions, traffic congestion, and potential obstacles. By analyzing these factors, DSS can assist in designing evacuation plans that ensure the safe and efficient movement of residents away from the fire. DSS can also assist in determining the appropriate timing of evacuation orders and the necessary resources to support evacuees, including shelters and transportation.

9.4.4 INCIDENT COORDINATION

Coordination among various agencies and teams is paramount for effective wildfire management. Decision Support Systems facilitate this coordination by providing a shared platform for communication and strategy development. DSS offer a centralized platform for all stakeholders involved in wildfire management, including firefighting teams, emergency services, and local authorities. This shared platform enables real-time communication, updates, and collaboration, ensuring that all parties are informed and working with the same information.

By providing a comprehensive view of the fire situation and available resources, DSS also support collaborative strategy development. Agencies can utilize the platform to develop and coordinate response strategies, share intelligence, and make joint decisions.

Decision Support Systems also facilitate real-time updates on fire conditions, resource status, and operational progress. This continuous flow of information ensures that all stakeholders are informed of the latest developments and can adjust their strategies accordingly.

9.4.5 PUBLIC COMMUNICATION

Public communication is a fundamental aspect of wildfire management, and Decision Support Systems can contribute to the dissemination of accurate and timely information to the community. DSS facilitate the generation of public alerts and warnings based on real-time data and predictive models. These systems can also automatically issue evacuation notices, safety warnings, and updates on fire conditions.

Furthermore, DSS can be integrated with various communication channels, such as social media platforms, emergency notification systems, and public service announcements. By providing accurate and transparent communication, DSS ensure that communities are adequately informed and prepared for potential impacts. This fosters community engagement through continuous communication and feedback.

9.5 INTERNET OF THINGS (IOT) AND SENSOR NETWORKS

Internet of Things (IoT) and sensor networks represent a transformative advancement in wildfire management by enabling real-time data collection, monitoring, and analysis (Giannakidou et al., 2024). These technologies integrate a diverse array of sensors and devices to provide continuous and precise data on environmental conditions and fire activity. This section delves into the role of Internet of Things (IoT) and sensor networks in wildfire management, specifically highlighting the contributions of environmental sensors, fire detection sensors, and networked systems.

9.5.1 ENVIRONMENTAL SENSORS

Environmental sensors integrated into Internet of Things (IoT) devices are pivotal in wildfire management, enabling the timely acquisition of real-time data on a diverse range of environmental parameters.

Temperature measurement: IoT devices equipped with temperature sensors continuously monitor ambient temperatures in various locations. Temperature data is crucial for assessing fire risk, as elevated temperatures can contribute to the likelihood of ignition and rapid fire spread. By tracking temperature variations, these sensors enable the identification of heat patterns and potential fire behavior.

Humidity monitoring: Humidity sensors measure the moisture content in the air, which significantly impacts fire risk and behavior. Low humidity levels enhance the probability of fire ignition and accelerate fire spread by drying out vegetation and fuels.

Wind speed and direction: Wind sensors gather data on wind speed and direction, which are pivotal factors influencing fire behavior. Wind can transport embers and flames to new areas, making it a key variable in predicting fire spread. Real-time wind data from IoT sensors facilitates fire movement forecasting and the development of effective firefighting strategies.

9.5.2 FIRE DETECTION SENSORS

Fire detection sensors are specialized IoT devices designed to identify the early signs of fire, such as smoke and heat, which are crucial for early warning and rapid response. One such system — InstaAlert — is depicted in figure 5.1.

IoT-enabled smoke detectors typically employ optical and ionization technologies to identify the presence of smoke. These detectors can be strategically deployed in diverse environments, such as remote locations, forests, and urban settings. Early detection of smoke enables prompt alerts to firefighting teams, potentially averting the development or spread of the fire.

Heat sensors detect elevated temperatures indicative of a fire. These sensors can pinpoint hot spots and unusual temperature surges that may signal the presence of a fire. By detecting heat at early stages, these sensors contribute to expedited response times and enhanced fire management.

Sensors are frequently integrated with early warning systems, ensuring that alerts are based on reliable and current information. This facilitates a more effective and coordinated response to fire incidents. Numerous fire detection sensors are specifically designed for remote operation, providing alerts and data even from isolated or challenging-to-reach areas. This capability expands the scope of fire detection and enhances overall wildfire surveillance.

9.5.3 NETWORKED SYSTEMS

The Internet of Things (IoT) framework establishes a robust network of sensors that continuously monitors environmental conditions and fire activity. Numerous interconnected sensors distributed across vast areas provide a continuous stream of data on temperature, humidity, wind, smoke, and heat. This uninterrupted data flow is fundamental for real-time monitoring and early detection of potential fire hazards.

The networked nature of IoT systems ensures that data is collected from multiple points, thereby enhancing the accuracy and timeliness of information. Data gathered by IoT sensors can be transmitted to central systems for analysis and integration with other data sources, such as remote sensing and Geographic Information

Systems (GIS). Seamless transmission and integration facilitate a comprehensive understanding of wildfire dynamics and support advanced modeling and prediction capabilities.

IoT networks can also be designed to adapt to changing conditions by deploying additional sensors or adjusting monitoring parameters based on real-time data. This adaptability ensures that the network remains responsive to evolving wildfire situations and environmental variations.

One of the most significant advantages of such IoT networks is their scalability and flexibility, enabling the expansion of sensor coverage as and when required. This scalability is particularly valuable in managing large or complex wildfire incidents, as additional sensors can be deployed to gather more data and enhance monitoring efforts.

9.6 ARTIFICIAL INTELLIGENCE (AI) AND MACHINE LEARNING (ML)

Artificial Intelligence (AI) and Machine Learning (ML) are revolutionizing wildfire management through advanced data analysis, enhanced predictive accuracy, and improved decision-making capabilities. These technologies harness large volumes of data to refine predictions, identify patterns, and quickly and continuously adapt to new information, ultimately supporting more effective wildfire management strategies.

9.6.1 FIRE BEHAVIOR PREDICTION

Artificial Intelligence (AI) models significantly enhance the prediction of fire behavior by analyzing extensive datasets to provide more accurate forecasts of fire spread and intensity. These models integrate a variety of data sources, including historical fire records, current weather conditions, and real-time sensor data. For instance, AI algorithms can process information on past fire incidents, weather patterns, and fuel types to predict how a fire will evolve under specific conditions. Advanced modeling techniques such as neural networks and ensemble methods enable AI to capture complex relationships between these variables, improving the accuracy of fire behavior predictions. As these models continuously learn from new data, they refine their forecasts over time, enabling more reliable predictions and better-informed decisions for firefighting strategies.

9.6.2 PATTERN RECOGNITION

Machine learning excels in detecting patterns and anomalies within vast datasets, which is crucial for identifying potential wildfire risks (Oliveira et al., 2021; Pais et al., 2021). Through the analysis of data gathered from environmental sensors, satellite imagery, and historical records, machine learning algorithms can discern anomalous patterns that may indicate heightened fire risk. For instance, a precipitous increase in temperature or a substantial alteration in vegetation vitality could serve as indicators of emerging fire hazards. This capacity facilitates the early identification of potential threats, thereby enabling prompt intervention and risk mitigation. Furthermore, machine learning algorithms can discern patterns associated with prior fire incidents to assess the likelihood of fire in novel areas. This predictive capability facilitates targeted risk management and assists in prioritizing resources where they are most indispensable.

9.6.3 ADAPTIVE LEARNING

Artificial Intelligence (AI) systems are meticulously designed to continuously acquire and adapt to novel data, thereby augmenting their efficacy over time. As these systems encounter fresh information, they revise their models and predictions, thereby enhancing their accuracy and reliability. This adaptive learning process entails feedback loops wherein the performance of predictions is juxtaposed against actual outcomes. By analyzing discrepancies and incorporating this feedback, AI systems refine their algorithms and augment their predictive capabilities. This continuous improvement facilitates superior decision-making by furnishing more precise forecasts and risk assessments. Ultimately, this contributes to more effective wildfire management strategies.

9.7 CLOUD COMPUTING

Cloud computing significantly contributes to wildfire management by providing scalable and flexible resources for data storage, processing, and immediate access. The majority of the maps included in this publication were generated through cloud computing processing, and systems such as SimplyFire operate exclusively on the cloud due to the multifaceted advantages offered by cloud computing over conventional computing.

9.7.1 DATA STORAGE AND PROCESSING

Given the exponential growth in data generation from remote sensing, IoT devices, and simulations, a robust system is necessary to host and process these vast quantities of data. The cost-prohibitive nature of individual high-performance computers and their limited utilization make shared computing resources indispensable. Cloud computing offers extensive storage and computational power, which are crucial for storing, managing, and processing large datasets. Scalable storage solutions enable cloud platforms to accommodate immense amounts of data. Cloud computing provides access to high-performance computing resources, facilitating the processing of intricate datasets and the execution of sophisticated fire behavior models. This capability supports comprehensive analysis and integration of data from diverse sources, leading to a comprehensive understanding of wildfire situations and enhanced decision-making.

9.7.2 REAL-TIME DATA ACCESS

One of the primary advantages of cloud computing is the provision of real-time data access. Cloud platforms facilitate immediate access to data and models from any location, which is crucial for timely decision-making in wildfire management. Fire managers and response teams can access up-to-date information on fire conditions, weather patterns, and resource availability from remote or field locations, without the need to visit their offices. This real-time access ensures that decisions are based on the most current data, thereby enhancing the effectiveness of response strategies. Furthermore, cloud platforms support continuous updates of data and models, providing real-time insights into fire behavior and conditions.

9.7.3 COLLABORATION

Cloud computing facilitates seamless collaboration among diverse stakeholders engaged in wildfire management by providing shared platforms for data and information exchange. Through cloud-based systems, multiple agencies, teams, and organizations can collaborate effectively by accessing and sharing data, models, and insights. This centralized platform enhances coordination and ensures that all stakeholders are aligned with the same information, eliminating the need for multiple separate copies of the same dataset. The capability to develop and implement joint strategies, share updates, and manage resources collectively is paramount for an effective wildfire response. Cloud computing also supports the dissemination of information to the public and other stakeholders, ensuring efficient communication of alerts, warnings, and updates.

10 Research and Development (R&D) in wildfire management

Research and Development (R&D) is crucial for advancing wildfire management strategies (Jazebi et al., 2019). Managing wildfires effectively has become increasingly crucial due to two primary reasons. Firstly, wildfires have become more frequent and intense due to evolving land use patterns and changing climate conditions. These wildfires, in turn, exacerbate climate change by releasing substantial quantities of greenhouse gases upon burning organic matter. Secondly, the remaining natural landscapes have become scarce, necessitating their protection to prevent their permanent extinction. Consequently, innovative solutions for prevention, detection, suppression, and recovery are increasingly necessary. This chapter delves into the latest advancements in research and development (R&D), encompassing improvements in fire behavior modeling, novel technologies, advancements in detection and monitoring systems, innovations in materials and vehicles, and R&D focused on enhancing the safety and efficiency of firefighters.

10.1 ADVANCEMENTS IN FIRE BEHAVIOR MODELING

Fire behavior modeling is central to understanding and predicting how fires spread under various conditions. Recent advancements have significantly improved the accuracy and reliability of these models (Hyde et al., 2012; Silva et al., 2022).

10.1.1 ENHANCED COMPUTATIONAL TECHNIQUES

Recent developments in computational techniques have led to more sophisticated fire behavior models. Machine learning (ML) and artificial intelligence (AI) have transformed fire modeling by analyzing vast datasets to predict fire behavior with ever greater precision (Bot and Borges, 2022; Jain et al., 2020). Machine learning algorithms analyze historical fire records, real-time sensor inputs, and weather forecasts to enhance the accuracy of fire spread and intensity predictions. Artificial intelligence models continuously acquire new data, enabling them to adapt to evolving fire dynamics and environmental factors.

10.1.2 INTEGRATION OF MULTI-SOURCE DATA

Contemporary fire behavior models now incorporate data from diverse sources, including satellite imagery, ground-based observations, and remote sensing technologies. By integrating these data sources, researchers can construct comprehensive models that incorporate variables (and their variations) such as vegetation types, fuel loads, and topography. This integration enhances prediction accuracy and facilitates informed decision-making for fire suppression and resource allocation.

10.1.3 REAL-TIME MODELING AND SIMULATION

Advancements in computing power and data integration have made real-time fire modeling and simulation increasingly feasible (Cardil et al., 2019; Wu et al., 2016). Real-time models provide timely and accurate predictions of fire behavior, enabling fire managers to respond swiftly and effectively to evolving circumstances. These models incorporate current data from weather stations, satellite imagery, and sensor networks to simulate fire propagation and forecast future behavior, thereby enhancing situational awareness and facilitating prompt decision-making.

10.2 DEVELOPMENT OF NEW TECHNOLOGIES

Technological innovation is paramount to enhancing wildfire management capabilities. Innovative tools and systems are being conceptualized to augment fire detection, monitoring, and response efforts.

10.2.1 ADVANCED SENSOR NETWORKS

Advanced sensor networks are at the forefront of wildfire detection and monitoring technologies (Dogra et al., 2021). Researchers are developing sensors such as thermal imaging cameras, smoke detectors, and weather stations to provide comprehensive data on fire conditions (Allison et al., 2016; Rahman et al., 2021; Vejmelka et al., 2016). These sensors can be deployed in remote and high-risk areas, providing real-time data on temperature, humidity, and smoke levels. Integrated sensor networks enable continuous monitoring and prompt detection of wildfires.

10.2.2 REMOTE SENSING TECHNOLOGIES

Remote sensing technologies, including satellites and drones, play a significant role in wildfire management. Satellites equipped with thermal and optical sensors provide large-scale observations of fire activity and environmental conditions. Drones equipped with thermal cameras and GPS systems offer high-resolution images and real-time data on fire behavior from aerial perspectives. These technologies enhance the ability to detect fires early, assess their progression, and guide firefighting efforts.

10.2.3 AUTONOMOUS SYSTEMS

Autonomous systems such as unmanned aerial vehicles (UAVs) and robots are being developed to assist in wildfire management (Couceiro et al., 2019; Ferreira et al.,

2020). Unmanned aerial vehicles can execute tasks such as fire mapping, surveillance, and data collection without endangering human operators. Robots specifically designed for tasks like clearing vegetation or creating firebreaks can operate in hazardous environments, thereby enhancing the efficiency and safety of wildfire response operations.

10.3 ENHANCEMENTS IN FIRE DETECTION AND MONITORING

Fire detection and monitoring systems are indispensable for effective wildfire management. Recent research and development endeavors are directed toward enhancing these systems to increase their accuracy and operational efficiency.

10.3.1 IMPROVED DETECTION ALGORITHMS

Research in fire detection algorithms aims to enhance the sensitivity and specificity of detection systems (Moayedi and Khasmakhi, 2023; Pereira et al., 2022). Advanced algorithms analyze data from diverse sensors and sources to discern early indications of fire, including alterations in temperature, smoke concentrations, and vegetation vitality. Enhanced detection algorithms diminish false alarms and augment the capacity to identify fires promptly, facilitating prompt responses.

10.3.2 INTEGRATION OF IOT AND CLOUD COMPUTING

The integration of Internet of Things (IoT) technology and cloud computing has transformed fire detection and monitoring systems (Bushnaq et al., 2021; Zope et al., 2020). Internet of Things devices, including smart sensors and connected cameras, provide real-time data on fire conditions. This data is processed and analyzed using cloud computing resources. This integration enables scalable and flexible monitoring systems that handle large volumes of data and offer real-time insights into fire behavior.

10.3.3 ENHANCED DATA VISUALIZATION AND ANALYSIS

Advancements in data visualization and analysis tools support better interpretation of fire data (Ahrens et al., 1997; Crawl et al., 2017; McCormick and Anrens, 1998). Advanced visualization platforms and analytical tools enable fire managers to effectively analyze intricate datasets, including fire progression maps, fuel load assessments, and weather forecasts. These tools facilitate a comprehensive understanding of fire dynamics, risk evaluation, and the formulation of well-informed decisions.

10.4 NEW MATERIALS

Advancements in materials science are significantly enhancing the effectiveness of wildfire prevention and suppression methodologies.

10.4.1 FIRE-RESISTANT MATERIALS

Recent scientific research has resulted in the development of advanced fire-resistant materials that are now incorporated into construction and landscaping practices. These materials, such as fire-resistant coatings, treated fabrics, and fire-resistant building components, offer enhanced protection for structures and infrastructure in areas prone to wildfires (Fedosov et al., 2020; Garver and Molander, 1962; Kilinc, 2013). These materials possess exceptional thermal resistance and ignition resistance, thereby mitigating the vulnerability of buildings and assets to wildfires.

10.4.2 FIRE-RESISTANT FOAMS AND GELS

Fire-resistant foams and gels enhance fire suppression efforts (Cui et al., 2019b; Harun-Ur-Rashid et al., 2023; Lyon et al., 1996). These materials can be applied to vegetation, structures, and other surfaces to establish a barrier that impedes the spread of fire and diminishes its intensity. Fire-resistant foams and gels are particularly valuable in regions where conventional fire suppression techniques are less effective, providing an additional tool for managing wildfires.

10.4.3 INNOVATIVE FIRE RETARDANTS

New fire retardants are being developed to slow down or stop the spread of fire by disrupting the combustion process. Advances in fire retardant technology include the creation of more environmentally friendly formulations that are less toxic and more effective (Horrocks and Price, 2008). Research also focuses on improving the delivery methods of fire retardants, such as using aerial application techniques to enhance coverage and efficiency (Yu et al., 2019).

10.5 MODERN VEHICLES

Managing wildfires in challenging terrains necessitates the utilization of specialized vehicles specifically engineered for rugged landscapes and limited access.

10.5.1 ALL-TERRAIN FIREFIGHTING VEHICLES

New all-terrain firefighting vehicles are being developed to combat wildfires in challenging and inaccessible areas. These vehicles are equipped with robust tires, elevated ground clearance, and sophisticated suspension systems to traverse rugged terrain. Equipped with water tanks, fire hoses, and other firefighting equipment, these vehicles can access remote locations and provide direct suppression support where conventional fire trucks are unable to reach.

10.5.2 AMPHIBIOUS FIREFIGHTING VEHICLES

Amphibious firefighting vehicles, which possess the ability to operate both on land and water, are gaining prominence in wildfire management, particularly in regions with aquatic sources such as lakes and rivers. These vehicles facilitate the transportation of water from these sources to the fire front, offering a versatile and efficient approach to firefighting in areas where water access is constrained. Their dual-environment capability enhances their effectiveness in a wide range of wildfire scenarios.

10.5.3 DRONE-BASED FIREFIGHTING SYSTEMS

Drone-based firefighting systems represent a cutting-edge innovation in wildfire management. Drones equipped with fire suppression tools, such as aerial water dispensers or fire retardant applicators, possess the capability to precisely target specific areas of a fire. These systems can be deployed in challenging-to-access locations, offering targeted assistance and enhancing the overall efficiency of firefighting operations.

10.6 FIREFIGHTING SAFETY AND EFFICIENCY

Ensuring the safety and efficiency of firefighters is a crucial aspect of R&D in wildfire management. Innovations in safety gear, equipment, and operational practices are essential for protecting firefighters and enhancing their effectiveness in the field (Koopmans et al., 2020; Navarro et al., 2022).

10.6.1 ADVANCED PERSONAL PROTECTIVE EQUIPMENT (PPE)

Research is focused on developing advanced personal protective equipment (PPE) for firefighters (Park et al., 2014; Santos et al., 2022). Innovations in personal protective equipment encompass heat-resistant fabrics, advanced breathing apparatuses, and improved communication systems. Contemporary PPE is meticulously engineered to offer enhanced protection against extreme temperatures, smoke inhalation, and physical hazards, while simultaneously prioritizing comfort and mobility. Enhanced breathing apparatuses with better filtration systems help protect firefighters from harmful smoke and toxins, reducing health risks (White and Hostler, 2017).

10.6.2 WEARABLE TECHNOLOGY

Wearable technology is becoming increasingly important in wildfire management. Research is exploring the use of smart sensors integrated into firefighter uniforms to monitor vital signs, such as heart rate and body temperature (Parker et al., 2017; Shakeriaski and Ghodrat, 2022). These sensors can provide real-time data on physiological stress and environmental conditions, enabling incident commanders to assess the well-being of firefighters and adjust operations accordingly. Additionally, wearable technology incorporates GPS devices for tracking and navigation, enhancing safety and coordination within the field.

10.6.3 FIREFIGHTING EQUIPMENT INNOVATIONS

Innovations in firefighting equipment prioritize enhancing efficiency and effectiveness. Research includes the development of more efficient fire hoses, advanced water delivery systems, and new firefighting agents (Ando et al., 2019; Han et al., 2017). For instance, lightweight and high-pressure hoses facilitate more efficient water delivery and facilitate handling. Advanced water delivery systems, such as those integrated with drones or remote-controlled vehicles, enhance the precision of water application, minimizing waste and optimizing suppression efforts.

10.6.4 ENHANCED TRAINING AND SIMULATION

Training and simulation technologies are crucial for preparing firefighters for real-world scenarios. Research focuses on developing advanced training tools, such as virtual reality (VR) and augmented reality (AR) systems, to simulate wildfire conditions and response scenarios (Narciso et al., 2020; Reis and Neves, 2019). These technologies offer immersive training environments that enable firefighters to practice tactics and decision-making in realistic settings, mitigating the risks associated with live training exercises. Advanced simulation tools enhance preparedness and ensure that firefighters are adequately equipped to handle intricate and dynamic fire scenarios.

10.7 INTEGRATING CLIMATE CHANGE

Climate change is significantly impacting the behavior and frequency of wildfires, necessitating substantial research and development efforts to mitigate these effects.

10.7.1 MODELING CLIMATE CHANGE IMPACTS

Research in climate change impacts focuses on understanding how changing climate conditions affect wildfire risks (An et al., 2015). Climate models integrated with fire behavior models provide insights into how variations in temperature, precipitation, and seasonal patterns may impact fire dynamics. This integration facilitates the development of adaptive management strategies that account for the evolving impacts of climate change.

10.7.2 DEVELOPING ADAPTIVE MANAGEMENT STRATEGIES

Adaptive management strategies are tailored to adapt to evolving conditions and emerging challenges. Research and development efforts focus on developing strategies that incorporate climate change projections and trends. For instance, fire management practices may be adjusted to account for anticipated changes in fire season duration, fuel availability, and fire intensity (Busenberg, 2004; Prudencio et al., 2018). These adaptive strategies enable fire managers to prepare for and respond to the evolving risks associated with a changing climate.

Section IV

Appendix

11 Wind movement

Wind refers to the large-scale movement of air, characterized by its speed and direction. Consequently, wind movement can be represented as a vector quantity, or arrow, whose length is directly proportional to the wind speed at a specific point, and whose direction corresponds to the direction of the wind. Wind speeds are measured using anemometers, and wind directions are measured using wind vanes. Winds are created by the net force acting on a parcel of air, which can be computed using the Bernoulli equation.

11.1 BERNOULLI EQUATION

For an incompressible non-viscous fluid in steady flow, the Bernoulli equation is written as:

$$\frac{1}{2}\rho v^2 + \rho gh + P = constant$$

where ρ is the density of the fluid, v is the speed of the fluid at any point, g is the acceleration due to gravity, h is the height of the fluid, and P is the pressure at any point in the fluid.

The equation essentially describes the conservation of energy. It states that the sum of kinetic energy, potential energy and pressure energy (or internal energy) in a fluid must be a constant. The equation is also commonly written as:

$$\frac{1}{2}v^2 + gh + \frac{P}{\rho} = constant$$

For a compressible fluid, such as an ideal gas, the equation is modified slightly:

$$\frac{1}{2}v^2 + gh + \frac{\gamma}{\gamma + 1}\frac{P}{\rho} = constant$$

where γ is the Laplace's coefficient for the ideal gas.

11.2 PRINCIPLES OF AIR MOVEMENT

The atmosphere is not an ideal gas nor is it non-viscous. Consequently, slight modifications to the Bernoulli equation are necessary when describing atmospheric flows. Nevertheless, certain fundamental principles can be elucidated:

1. Air flows from regions of high pressure to regions of low pressure.

2. Denser air descends due to gravity, while less dense air ascends.
3. The density of air is influenced by various factors, including temperature (as air at elevated temperatures expands and consequently experiences a reduction in density) and composition (the addition of heavier gases, such as carbon dioxide, enhances the density of air).
4. Air movement is modified by topography (e.g. mountains can slow down winds or cause them to move up the slope), and various forces including frictional, viscous, and Coriolis force.

11.3 GENERAL CIRCULATION

The general circulation of the atmosphere is governed by several factors, including:

1. the latitudinal differences in energy surplus and deficit, making areas near the equator energy-rich and hot, and areas near the poles energy-deficient and cold,
2. the emergence of pressure belts when air in hotter areas moves up through convection creating low-pressure zones, and areas where this air subsides becoming high-pressure zones,
3. the migration of pressure belts with the seasons due to the tilt of the Earth's axis and its revolution around the Sun,
4. the distribution of continents and oceans which not only create a land-rich Northern hemisphere and an ocean-rich Southern hemisphere, but also local variations,
5. the rotation of the Earth, which creates the Coriolis force, and
6. the ocean currents, which also contribute to the movement of energy on the planet.

These lead to the formation of various 'cells' [Figure 11.1a]:

1. **Hadley Cell**: The Hadley cell is a prominent atmospheric circulation pattern that exists between the equator and 30 degrees North and South latitudes. The region near the equator experiences the highest insolation (solar radiation reaching a given area), leading to the heating of the air and subsequent reduction in its density. Consequently, the air rises, creating a low-pressure zone. As the air ascends, it cools and its density increases. Approximately at the 30th parallel, the air descends, forming a high-pressure area. This pattern results in the presence of high-pressure areas near the 30th parallels and low-pressure areas near the equator. The movement of air from high-pressure zones to low-pressure zones generates winds. The Coriolis force deflects these winds to the westward direc-

tion. These west-moving winds are known as the Trade winds (so-called due to their use in sail ships facilitating trade) or Easterlies (since they come from the East). Trade winds are permanent prevailing winds that exhibit seasonal variations in their intensity. The region where the trade winds of Northern and Southern hemispheres meet is known as the Intertropical Convergence Zone (ITCZ), or the doldrums (from *dull*, since the area often had calm weather, making it dull from the point of view of sail ships, though interspersed with sudden storms and light, unpredictable winds). Although situated near the equator, the Intertropical Convergence Zone (ITCZ) experiences a shift to the Northern Hemisphere around June and July and to the Southern Hemisphere around December and January. This movement is directly correlated with the apparent North-South shift of the Sun as the seasons change.

2. **Polar Cell**: The air near the 60th parallel experiences a temperature gradient, being warmer than the polar air. Consequently, this air rises and undergoes movement in both poleward and equatorward directions. The portion moving towards the poles cools, increases in density, and descends, creating a high-pressure area near the poles known as the 'polar highs'. Surface winds are generated from this high-pressure area near the poles towards the low-pressure area near the 60th parallel, turning westwards due to the Coriolis force. These winds are known as the 'polar easterlies'. The polar cell is the smallest and weakest of the three cells.

3. **Ferrel Cell**: Part of the rising air near the 60th parallel moves towards the equator and collides with the high-level air of the Hadley cell. It descends near the 30th parallel, substantiating the high-pressure zone. Surface winds move from this high-pressure area near the 30th parallel towards the low-pressure area near the 60th parallel, turning eastwards due to the Coriolis force. This east-turned wind appears to be coming from the west direction and is called 'prevailing westerlies'.

11.4 LOCAL WINDS

Besides the general circulation in the atmosphere, we have several local winds. Local winds are formed:

1. on a small spatial scale (tens to a few hundred kilometers in horizontal extent),

2. for a short period of time (typically few hours to few days), and

3. due to local conditions.

They are classified into periodical and non-periodical local winds:

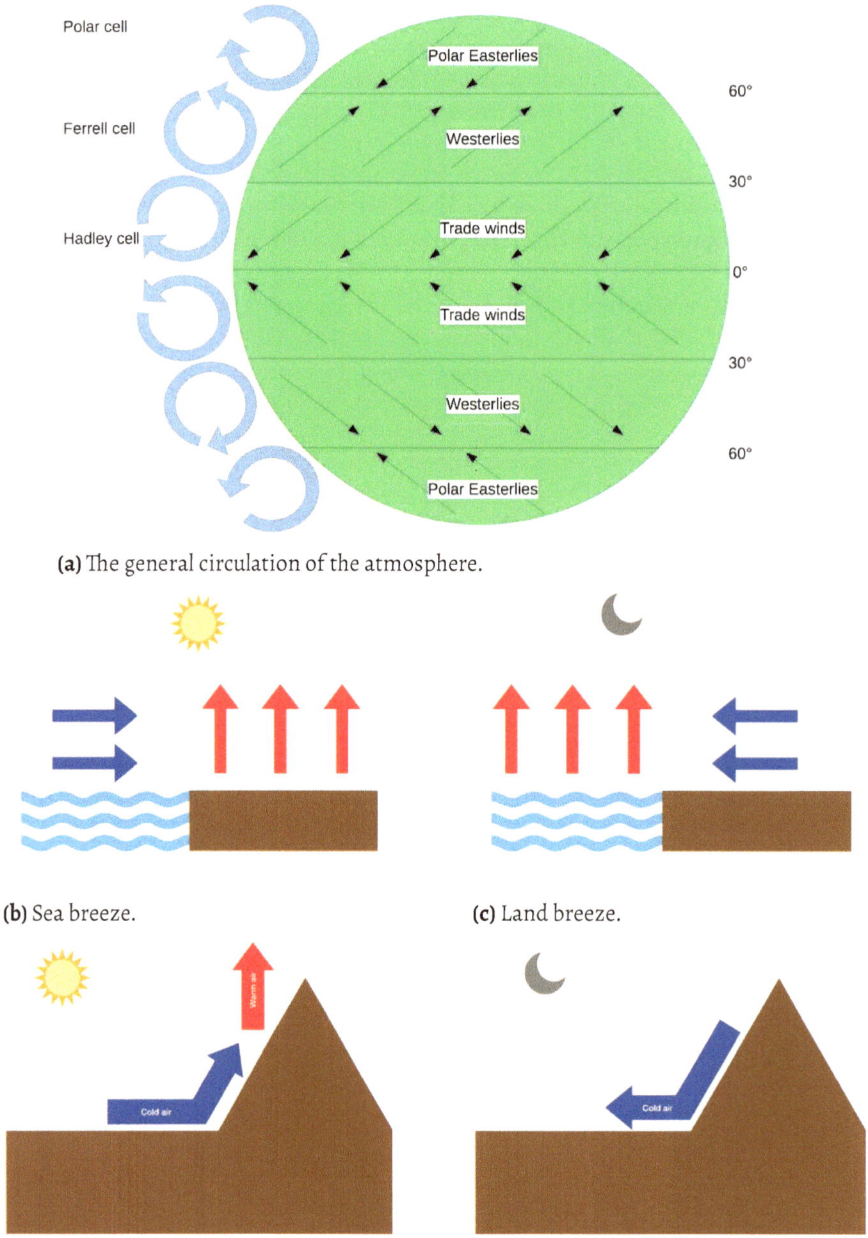

(a) The general circulation of the atmosphere.

(b) Sea breeze.

(c) Land breeze.

(d) Valley wind.

(e) Mountain wind.

FIGURE 11.1: Some wind movements in the Earth's atmosphere.

11.4.1 PERIODICAL WINDS

These winds originate and occur at a set pattern, generally showing a day-night variation. Good examples include the sea and the land breeze. During daylight hours, the Sun's energy is absorbed more rapidly and extensively by land compared to water. This phenomenon is attributed to several factors, including land's darker appearance, which enhances its ability to absorb incoming radiation, and its poor heat conduction properties. Consequently, the upper layers of land retain most of the incoming energy, resulting in a significant temperature rise. In contrast, water possesses a high heat capacity and cools through evaporation, maintaining a cooler water surface temperature. The differential temperature between land and water generates a local wind. The air above the land surface heats up, becomes less dense, and ascends. Conversely, the air above the water surface remains cooler and denser. This cool, dense air rushes to fill the void created by the convection current above the land surface, thereby establishing a local wind. This wind, coming from the sea, is called a sea breeze [Figure 11.1b].

During the nighttime, a reverse wind movement occurs. The darker-colored land, being an effective heat radiator, releases the stored energy from its surface, resulting in a cold surface. In contrast, the water surface, with its high heat capacity and light color, retains much warmer temperatures due to its poor ability to radiate energy. The air in contact with the water surface becomes heated and less dense, causing it to rise. Conversely, the cool and dense air over the land rushes to fill the void created by the convection current above the water surface, thereby generating a local wind. This wind, coming from the land, is called a land breeze [Figure 11.1c].

A comparable process occurs in mountainous regions. During the daytime, mountain slopes face the Sun and absorb its energy, resulting in significant warming compared to the surrounding valley areas. The air in contact with the heated surface becomes warmer and less dense, rising upward. Conversely, the cooler air in contact with the valley regions ascends to fill the void created by the convection current above the mountainous surface, thereby generating a local wind. This wind, coming from the valley region, is called a valley wind [Figure 11.1d]. It is important to note that in this case, the wind is moving upwards, towards a region of low pressure (since atmospheric pressure decreases with height). When it does so, each air parcel expands and cools in a process known as adiabatic cooling (since the process of cooling happens only due to expansion, and not because of any transfer of energy from the air to its surroundings). As the air cools, its relative humidity increases. Thus, valley winds often feel cold and wet.

During the nighttime, mountain slopes radiate away their stored energy, resulting in a significant drop in temperature compared to the surrounding valley areas. The air in contact with the cold surface cools and becomes denser, descending towards the valley regions and generating a local wind. This wind, coming from the mountain region, is called a mountain wind [Figure 11.1e]. It is important to note that in this case, the wind is moving downwards, towards a region of higher pressure (since atmospheric pressure increases with descent). When it does so, each air parcel contracts and warms in a process known as adiabatic heating (since the process of heating happens only due to contraction, and not because of any transfer of energy to the air from its surroundings). As the air warms, its relative humidity decreases. Thus, mountain winds often feel hot and dry. These winds are also called descent winds or katabatic winds (from Greek *katabasis* = descent). In the case of the wildfire that burnt Paradise City, there was a strong katabatic wind movement towards the low-lying areas [Figure 1.1c–d] in the early morning time when the phenomenon was at its strongest, with the mountain slopes having had all night to radiate energy. These winds, locally called Jarbo winds, being hot and dry, greatly accentuated and accelerated the spread of wildfire and were a main reason for the massive destruction caused by the wildfire.

11.4.2 NON-PERIODICAL WINDS

These winds are present during a season (often due to adiabatic heating and cooling, or passage over hot or cold areas), and are further classified into

- hot winds, such as Chinook, Santa Ana, and loo, and
- cold winds, such as mistral, Northers, and Levanter.

11.5 OTHER WIND PHENOMENA

Apart from prevailing and local winds, we also have special phenomena such as monsoons and cyclones. While not very important for the spread of wildfires, these often bring with them rainfall that proves instrumental in containing the spread of wildfires.

References

Acea, M. J. and Carballas, T. (1996). Changes in physiological groups of microorganisms in soil following wildfire. *FEMS Microbiology Ecology*, 20(1):33–39.

Adkins, A. P., Bowman, T., Cerasoli, M., Sharafati, C., Short, J., Lewis, K., Martinez, W., Pierce, K., Titus, C., and Young, S. (2021). Aerial data acquisition platform for firefront monitoring.

Aedo, S. A. and Bonilla, C. A. (2021). A numerical model for linking soil organic matter decay and wildfire severity. *Ecological Modelling*, 447:109506.

Afghah, F., Razi, A., Chakareski, J., and Ashdown, J. (2019). Wildfire monitoring in remote areas using autonomous unmanned aerial vehicles. In *IEEE INFOCOM 2019-IEEE Conference on Computer Communications Workshops (INFOCOM WKSHPS)*, pages 835–840. IEEE.

Agency, F. E. M. (2017). *National incident management system*. FEMA.

Ager, A. A., Finney, M. A., Kerns, B. K., and Maffei, H. (2007). Modeling wildfire risk to northern spotted owl (strix occidentalis caurina) habitat in central oregon, usa. *Forest Ecology and Management*, 246(1):45–56.

Ahrens, J., McCormick, P., Bossert, J., Reisner, J., and Winterkamp, J. (1997). Case study: Wildfire visualization. In *Proceedings. Visualization'97 (Cat. No. 97CB36155)*, pages 451–454. IEEE.

Albini, F. A. (1976a). *Computer-based models of wildland fire behavior: a user's manual*. Intermountain Forest and Range Experiment Station, Forest Service, US Department of Agriculture.

Albini, F. A. (1976b). *Estimating wildfire behavior and effects*, volume 30. Department of Agriculture, Forest Service, Intermountain Forest and Range Experiment Station.

Alharbi, O. M., Khattab, R. A., Ali, I., et al. (2018). Health and environmental effects of persistent organic pollutants. *Journal of Molecular Liquids*, 263:442–453.

Ali, H., Khan, E., and Sajad, M. A. (2013). Phytoremediation of heavy metals—concepts and applications. *Chemosphere*, 91(7):869–881.

Allison, R. S., Johnston, J. M., Craig, G., and Jennings, S. (2016). Airborne optical and thermal remote sensing for wildfire detection and monitoring. *Sensors*, 16(8):1310.

Ambrosia, V. G., Buechel, S. W., Brass, J. A., Peterson, J. R., Davies, R. H., Kane, R. J., and Spain, S. (1998). An integration of remote sensing, gis, and information distribution for wildfire detection and management. *Photogrammetric Engineering and Remote Sensing*, 64:977–986.

An, H., Gan, J., and Cho, S. J. (2015). Assessing climate change impacts on wildfire risk in the united states. *Forests*, 6(9):3197–3211.

Anderson, P. (1995). Ecological restoration and creation: a review. *Biological Journal of the Linnean Society*, 56(suppl_1):187–211.

Ando, H., Ambe, Y., Yamaguchi, T., Konyo, M., Tadakuma, K., Maruyama, S., and Tadokoro, S. (2019). Fire fighting tactics with aerial hose-type robot "dragon firefighter". In *2019 IEEE International Conference on Advanced Robotics and its Social Impacts (ARSO)*, pages 291–297. IEEE.

Andreae, M. O. (2019). Emission of trace gases and aerosols from biomass burning–an updated assessment. *Atmospheric Chemistry and Physics*, 19(13):8523–8546.

Andreae, M. O. and Merlet, P. (2001). Emission of trace gases and aerosols from biomass burning. *Global biogeochemical cycles*, 15(4):955–966.

Antonini, E., Brunori, M., Rotilio, G. C., Greenwood, C., and MalmstrÖM, B. G. (1971). The interaction of cyanide with cytochrome oxidase. *European journal of biochemistry*, 23(2):396–400.

Apfelbaum, S. and Haney, A. (1981). Bird populations before and after wildfire in a great lakes pine forest. *The Condor*, 83(4):347–354.

Armstrong, A., Benn, G., Bowland, A., Goodman, P., Johnson, D., Maddock, A., and Scott-Shaw, C. (1998). Plantation forestry in south africa and its impact on biodiversity. *Southern African Forestry Journal*, 182(1):59–65.

Artés, T., Oom, D., De Rigo, D., Durrant, T. H., Maianti, P., Libertà, G., and San-Miguel-Ayanz, J. (2019). A global wildfire dataset for the analysis of fire regimes and fire behaviour. *Scientific data*, 6(1):296.

Ascoli, D., Russo, L., Giannino, F., Siettos, C., and Moreira, F. (2020). Firebreak and fuelbreak. In *Encyclopedia of wildfires and wildland-urban interface (WUI) fires*, pages 444–452. Springer.

Athanasis, N., Karagiannis, F., Palaiologou, P., Vasilakos, C., and Kalabokidis, K. (2015). Aegis app: wildfire information management for windows phone devices. *Procedia Computer Science*, 56:544–549.

Aubry-Wake, C., Bertoncini, A., and Pomeroy, J. W. (2022). Fire and ice: The impact of wildfire-affected albedo and irradiance on glacier melt. *Earth's Future*, 10(4):e2022EF002685.

Authority, E. F. S. (2020). Outcome of the consultation with member states and efsa on the basic substance application for approval of dimethyl sulfide to be used in plant protection as a non-lethal food attractant for truffle beetle. Technical report, Wiley Online Library.

Awadhiya, A. (2016). *RS-GIS with free software*. Indira Gandhi National Forest Academy, Dehradun.

Awadhiya, A. (2017). Fire breakers: novel devices for the control of forest surface fires. *Indian Forester*, 143.

Awadhiya, A. (2018). Generation of division level fire maps using satellite-derived fire data. *Indian Forester*, 144.

Awadhiya, A. (2021). *Principles of Wildlife Conservation*. CRC Press.

Awadhiya, A. (2022a). *Burn severity atlas of Madhya Pradesh*. Madhya Pradesh Forest Department.

Awadhiya, A. (2022b). *Madhya Pradesh Forest Fire Compendium*. Madhya Pradesh Forest Department.

Aydin, B., Selvi, E., Tao, J., and Starek, M. J. (2019). Use of fire-extinguishing balls for a conceptual system of drone-assisted wildfire fighting. *Drones*, 3(1):17.

Banerjee, S., Chakraborty, A., Armenteras, D., Dutta, D., Kundu, A., and Datta, D. (2024). Vegetation, soil, and livelihoods: The complex effects of forest fires on eastern india's dry deciduous forests. *Land Degradation & Development*.

Banks, S. C., Dujardin, M., McBurney, L., Blair, D., Barker, M., and Lindenmayer, D. B. (2011). Starting points for small mammal population recovery after wildfire: recolonisation or residual populations? *Oikos*, 120(1):26–37.

Barbier, G., Zafarani, R., Gao, H., Fung, G., and Liu, H. (2012). Maximizing benefits from crowdsourced data. *Computational and Mathematical Organization Theory*, 18:257–279.

Barker, R. (2013). *Scorched earth: How the fires of Yellowstone changed America*. Island Press.

Bartels, S. F., Chen, H. Y., Wulder, M. A., and White, J. C. (2016). Trends in post-disturbance recovery rates of canada's forests following wildfire and harvest. *Forest Ecology and Management*, 361:194–207.

Bartholomaeus, A. R. and Haritos, V. S. (2005). Review of the toxicology of carbonyl sulfide, a new grain fumigant. *Food and chemical toxicology*, 43(12):1687–1701.

Bates, F. L. and Peacock, W. G. (1989). Long term recovery. *International Journal of Mass Emergencies & Disasters*, 7(3):349–365.

Bayham, J., Yoder, J. K., Champ, P. A., and Calkin, D. E. (2022). The economics of wildfire in the united states. *Annual Review of Resource Economics*, 14:379–401.

Beil, K. M. (1999). *Fire in their eyes: Wildfires and the people who fight them*. Houghton Mifflin Harcourt.

Benedict, K. B., Prenni, A. J., Carrico, C. M., Sullivan, A. P., Schichtel, B. A., and Collett Jr, J. L. (2017). Enhanced concentrations of reactive nitrogen species in wildfire smoke. *Atmospheric Environment*, 148:8–15.

Berryoung, M. (2021). Little boxes on the hillside: An inventory and analysis of wildfire lookout structures in the pacific northwest region 06 of the united states forest service. Master's thesis, University of Oregon.

Bibby, C. J. (1978). Conservation of the dartford warbler on english lowland heaths: a review. *Biological Conservation*, 13(4):299–307.

Bodí, M. B., Martin, D. A., Balfour, V. N., Santín, C., Doerr, S. H., Pereira, P., Cerdà, A., and Mataix-Solera, J. (2014). Wildland fire ash: production, composition and eco-hydro-geomorphic effects. *Earth-Science Reviews*, 130:103–127.

Bogdos, N. and Manolakos, E. S. (2019). Crowd-sourced wildfire spread prediction with remote georeferencing using smartphones. *IEEE Access*, 7:102102–102112.

Bond, M. and Bradley, C. (2003). Impacts of the 2003 southern california wildfires on four species listed as threatened or endangered under the federal endangered species act. *Center for Biological Diversity. Unpublished Report.*

Bond, W. J. and Van Wilgen, B. W. (2012). *Fire and plants*, volume 14. Springer Science & Business Media.

Borgogni, F., Lavecchia, A., Mastrolonardo, G., Certini, G., Ceccherini, M. T., and Pietramellara, G. (2019). Immediate-and short-term wildfire impact on soil microbial diversity and activity in a mediterranean forest soil. *Soil Science*, 184(2):35–42.

Bosso, L., Ancillotto, L., Smeraldo, S., D'Arco, S., Migliozzi, A., Conti, P., and Russo, D. (2018). Loss of potential bat habitat following a severe wildfire: a model-based rapid assessment. *International Journal of Wildland Fire*, 27(11):756–769.

Bot, K. and Borges, J. G. (2022). A systematic review of applications of machine learning techniques for wildfire management decision support. *Inventions*, 7(1):15.

Bowman, D. (2018). Wildfire science is at a loss for comprehensive data. *Nature*, 560(7716):7–8.

Bramwell, L. (2008). Hotshots: The origins and work culture of america's elite wildland firefighters. *New Mexico Historical Review*, 83(3):2.

Brehme, C. S., Clark, D. R., Rochester, C. J., and Fisher, R. N. (2011). Wildfires alter rodent community structure across four vegetation types in southern california, usa. *Fire Ecology*, 7(2):81–98.

Brenkert-Smith, H., Dickinson, K. L., Champ, P. A., and Flores, N. (2013). Social amplification of wildfire risk: the role of social interactions and information sources. *Risk analysis*, 33(5):800–817.

Brodie, E. G., Knapp, E. E., Brooks, W. R., Drury, S. A., and Ritchie, M. W. (2024). Forest thinning and prescribed burning treatments reduce wildfire severity and buffer the impacts of severe fire weather. *Fire Ecology*, 20(1):17.

Brooker, M. G. and Rowley, I. C. R. (1991). Impact of wildfire on the nesting behaviour of birds in heathland. *Wildlife Research*, 18:249–263.

Brown, D. J., Mali, I., and Forstner, M. R. J. (2014). Wildfire and postfire restoration action effects on microclimate and seedling pine tree survivorship. *Journal of Fish and Wildlife Management*, 5:174–182.

Budisulistiorini, S. H., Riva, M., Williams, M., Miyakawa, T., Chen, J., Itoh, M., Surratt, J. D., and Kuwata, M. (2018). Dominant contribution of oxygenated organic aerosol to haze particles from real-time observation in singapore during an indonesian wildfire event in 2015. *Atmospheric Chemistry and Physics*, 18(22):16481–16498.

Burgan, R. E. (1984). *Behave: fire behavior prediction and fuel modeling system, fuel subsystem*, volume 167. US Department of Agriculture, Forest Service, Intermountain Forest and Range.

Buscemi, D., Hoffman, D., Vyas, N., Spann, J., and Kuenzel, W. (2002). Effects of phos-chek g75-f and silv-ex on developing northern bobwhite quail (colinus virginianus). *Archives of environmental contamination and toxicology*, 43:0330–0337.

Busenberg, G. J. (2004). Adaptive policy design for the management of wildfire hazards. *American Behavioral Scientist*, 48(3):314–326.

Bushnaq, O. M., Chaaban, A., and Al-Naffouri, T. Y. (2021). The role of uav-iot networks in future wildfire detection. *IEEE Internet of Things Journal*, 8(23):16984–16999.

Butcher, K. (2019). Wildfire exposure to critical habitat of endangered and threatened species in california. Master's thesis, University of San Francisco.

Byard, R. (2011). Issues in the classification and pathological diagnosis of asphyxia. *Australian Journal of Forensic Sciences*, 43(1):27–38.

Byard, R. W. (2023). Lethal asphyxia. In *Forensic and Legal Medicine*, pages 287–297. CRC Press.

Cadd, T. L., Slyman, P. M., and Maizer, L. M. (1999). The ics planning process exercise: A new exercise tool. In *International Oil Spill Conference*, volume 1999, pages 411–415. American Petroleum Institute.

Cameron, P. A., Mitra, B., Fitzgerald, M., Scheinkestel, C. D., Stripp, A., Batey, C., Niggemeyer, L., Truesdale, M., Holman, P., Mehra, R., et al. (2009). Black saturday: the immediate impact of the february 2009 bushfires in victoria, australia. *Medical Journal of Australia*, 191(1):11–16.

Carballo-Leyenda, B., Villa, J. G., López-Satué, J., and Rodríguez-Marroyo, J. A. (2017). Impact of different personal protective clothing on wildland firefighters' physiological strain. *Frontiers in physiology*, 8:618.

Cardil, A., Monedero, S., Silva, C. A., and Ramirez, J. (2019). Adjusting the rate of spread of fire simulations in real-time. *Ecological Modelling*, 395:39–44.

Center, N. G. S. F. (2013a). Bandpassesl7vl8-jul2013.jpg. https://landsat.gsfc.nasa.gov/wp-content/uploads/2013/01/BandpassesL7vL8-Jul20131.pdf. Accessed: (2024-03-05).

Center, N. G. S. F. (2013b). Wavelength, frequency, and energy. https://imagine.gsfc.nasa.gov/science/toolbox/spectrum_chart.html. Accessed: (2024-03-05).

Certini, G., Moya, D., Lucas-Borja, M. E., and Mastrolonardo, G. (2021). The impact of fire on soil-dwelling biota: A review. *Forest Ecology and Management*, 488:118989.

Certini, G., Nocentini, C., Knicker, H., Arfaioli, P., and Rumpel, C. (2011). Wildfire effects on soil organic matter quantity and quality in two fire-prone mediterranean pine forests. *Geoderma*, 167:148–155.

Çetin, A. E., Dimitropoulos, K., Gouverneur, B., Grammalidis, N., Günay, O., Habiboğlu, Y. H., Töreyin, B. U., and Verstockt, S. (2013). Video fire detection–review. *Digital Signal Processing*, 23(6):1827–1843.

Chandra, K. (2006). *Faunal Resources of National Parks of Madhya Pradesh and Chhattisgarh*. Zoological Survey of India.

Chang, H.-H. (2017). A literature review and analysis of the incident command system. *International journal of emergency management*, 13(1):50–67.

Chas-Amil, M. L., Prestemon, J. P., McClean, C. J., and Touza, J. (2015). Human-ignited wildfire patterns and responses to policy shifts. *Applied Geography*, 56:164–176.

Chauhan, A. and Hughes, A. L. (2017). Providing online crisis information: An analysis of official sources during the 2014 carlton complex wildfire. In *Proceedings of the 2017 CHI Conference on Human Factors in Computing Systems*, pages 3151–3162.

Chen, H., Uzun, H., Chow, A. T., and Karanfil, T. (2020a). Low water treatability efficiency of wildfire-induced dissolved organic matter and disinfection by-product precursors. *Water Research*, 184:116111.

Chen, J., Pangle, L. A., Gannon, J. P., and Stewart, R. D. (2020b). Soil water repellency after wildfires in the blue ridge mountains, united states. *International Journal of Wildland Fire*, 29(11):1009–1020.

Chief, K., Young, M. H., and Shafer, D. S. (2012). Changes in soil structure and hydraulic properties in a wooded-shrubland ecosystem following a prescribed fire. *Soil Science Society of America Journal*, 76(6):1965–1977.

Chittenden, H. M. (2018). *The Yellowstone National Park*. BoD–Books on Demand.

Choi, J., Oh, J. Y., Lee, Y. S., Min, K. H., Hur, G. Y., Lee, S. Y., Kang, K. H., and Shim, J. J. (2018). Harmful impact of air pollution on severe acute exacerbation of chronic obstructive pulmonary disease: particulate matter is hazardous. *International journal of chronic obstructive pulmonary disease*, pages 1053–1059.

Chouksey, S. and Singh, S. (2018). Assessments on the impact of human-tiger conflict and community-based conservation in bandhavgarh tiger reserve, madhya pradesh, india. *Journal of Threatened Taxa*, 10(7):11844–11849.

Christensen, N. L., Agee, J. K., Brussard, P. F., Hughes, J., Knight, D. H., Minshall, G. W., Peek, J. M., Pyne, S. J., Swanson, F. J., Thomas, J. W., et al. (1989). Interpreting the yellowstone fires of 1988. *BioScience*, 39(10):678–685.

Chuvieco, E., Aguado, I., Salas, J., García, M., Yebra, M., and Oliva, P. (2020). Satellite remote sensing contributions to wildland fire science and management. *Current Forestry Reports*, 6:81–96.

Colorado, H. A. and Henkin, J. M. (2023). Fire-resistant plants: A review of plant morphology, tissues, habits, ecological adaptations, and other factors contributing to bioderived environmental solutions and technologies. *Engineered Science*, 27:1024.

Couceiro, M. S., Portugal, D., Ferreira, J. F., and Rocha, R. P. (2019). Semfire: Towards a new generation of forestry maintenance multi-robot systems. In *2019 IEEE/SICE International Symposium on System Integration (SII)*, pages 270–276. IEEE.

Crampton, J. W., Graham, M., Poorthuis, A., Shelton, T., Stephens, M., Wilson, M. W., and Zook, M. (2013). Beyond the geotag: situating 'big data'and leveraging the potential of the geoweb. *Cartography and geographic information science*, 40(2):130–139.

Crawl, D., Block, J., Lin, K., and Altintas, I. (2017). Firemap: A dynamic data-driven predictive wildfire modeling and visualization environment. *Procedia Computer Science*, 108:2230–2239.

Crow, D. A., Lawhon, L. A., Koebele, E., Kroepsch, A., Schild, R., and Huda, J. (2015). Information, resources, and management priorities: agency outreach and mitigation of wildfire risk in the western united states. *Risk, Hazards & Crisis in Public Policy*, 6(1):69–90.

Cuevas-Gonzalez, M., Gerard, F., Balzter, H., and Riano, D. (2009). Analysing forest recovery after wildfire disturbance in boreal siberia using remotely sensed vegetation indices. *Global Change Biology*, 15(3):561–577.

Cui, X., Alam, M. A., Perry, G. L., Paterson, A. M., Wyse, S. V., and Curran, T. J. (2019a). Green firebreaks as a management tool for wildfires: Lessons from china. *Journal of environmental management*, 233:329–336.

Cui, X. F., Zheng, W. J., Zou, W., Liu, X. Y., Yang, H., Yan, J., and Gao, Y. (2019b). Water-retaining, tough and self-healing hydrogels and their uses as fire-resistant materials. *Polymer Chemistry*, 10(37):5151–5158.

Dalezios, N. R., Kalabokidis, K., Koutsias, N., and Vasilakos, C. (2017). Wildfires and remote sensing: An overview. *Remote Sensing of Hydrometeorological Hazards*, pages 211–236.

David, A. T., Asarian, J. E., and Lake, F. K. (2018). Wildfire smoke cools summer river and stream water temperatures. *Water Resources Research*, 54(10):7273–7290.

Davies, D. K., Ilavajhala, S., Wong, M. M., and Justice, C. O. (2008). Fire information for resource management system: archiving and distributing modis active fire data. *IEEE Transactions on Geoscience and Remote Sensing*, 47(1):72–79.

Davies, G., Bakker, J., Dettweiler-Robinson, E., Dunwiddie, P. W., Hall, S., Downs, J., and Evans, J. (2012). Trajectories of change in sagebrush steppe vegetation communities in relation to multiple wildfires. *Ecological Applications*, 22(5):1562–1577.

Dawe, D. A., Parisien, M., Dongen, A. V., and Whitman, E. (2021). Initial succession after wildfire in dry boreal forests of northwestern north america. *Plant Ecology*, 223:789 – 809.

Deal, T., Huyck, V., Merrick, G., Mills, C., and de Bettencourt, M. (2006). *Beyond initial Response: Using the national incident management system's incident command system*. AuthorHouse.

Debano, L. F. and Krammes, J. (1966). Water repellent soils and their relation to wildfire temperatures. *Hydrological Sciences Journal*, 11(2):14–19.

Deeming, J. E. (1972). *National fire-danger rating system*, volume 84. Rocky Mountain Forest and Range Experiment Station, Forest Service, US Department of Agriculture.

Dennis, F. C. et al. (1999). *Fire-resistant landscaping*. Colorado State University Cooperative Extension Fort Collins, CO, USA.

Dennison, P. E., Charoensiri, K., Roberts, D. A., Peterson, S. H., and Green, R. O. (2006). Wildfire temperature and land cover modeling using hyperspectral data. *Remote Sensing of Environment*, 100(2):212–222.

Desmond, M. (2006). Becoming a firefighter. *Ethnography*, 7(4):387–421.

Detweiler, A. J., Fitzgerald, S. A., et al. (2006). Fire-resistant plants for home landscapes: Selecting plants that may reduce your risk from wildfire. Technical report, Oregon State University Extension Service.

Dickerson, R., Kondragunta, S., Stenchikov, G., Civerolo, K., Doddridge, B., and Holben, B. (1997). The impact of aerosols on solar ultraviolet radiation and photochemical smog. *science*, 278(5339):827–830.

Dietrich, J. P., Van Gaest, A. L., Strickland, S. A., Hutchinson, G. P., Krupkin, A. B., and Arkoosh, M. R. (2014). Toxicity of phos-chek lc-95a and 259f fire retardants to ocean-and stream-type chinook salmon and their potential to recover before seawater entry. *Science of the total environment*, 490:610–621.

Doermann, J. L., Kuligowski, E. D., and Milke, J. (2021). From social science research to engineering practice: Development of a short message creation tool for wildfire emergencies. *Fire Technology*, 57:815–837.

Doerr, S., Shakesby, R., Blake, W., Chafer, C., Humphreys, G., and Wallbrink, P. (2006). Effects of differing wildfire severities on soil wettability and implications for hydrological response. *Journal of Hydrology*, 319(1-4):295–311.

Dogra, R., Rani, S., and Sharma, B. (2021). A review to forest fires and its detection techniques using wireless sensor network. In *Advances in communication and computational technology: select proceedings of ICACCT 2019*, pages 1339–1350. Springer.

Donovan, V. M., Roberts, C. P., Fogarty, D. T., Wedin, D. A., and Twidwell, D. (2022). Targeted grazing and mechanical thinning enhance forest stand resilience under a narrow range of wildfire scenarios. *Ecosphere*, 13(5):e4061.

Dorrer, G., Dorrer, A., Buslov, I., and Yarovoy, S. (2018). System of personnel training in decision making in fighting wildfires. In *IOP Conference Series: Materials Science and Engineering*, volume 450, page 062018. IOP Publishing.

Dubey, S., Pant, H., and Malviya, S. J. (2021). Avian fauna of bandhavgarh national park, madhya pradesh. *Journal of Natural Resource and Development*, 16(1).

Duff, R., Grant, L. D., and van den Heever, S. C. (2014). Effects of wildfire pollution on the microphysical and electrical properties of pyrocumulus. In *AGU Fall Meeting Abstracts*, volume 2014, pages A21I–3143.

Dupuy, J.-l., Fargeon, H., Martin-StPaul, N., Pimont, F., Ruffault, J., Guijarro, M., Hernando, C., Madrigal, J., and Fernandes, P. (2020). Climate change impact on future wildfire danger and activity in southern europe: a review. *Annals of Forest Science*, 77:1–24.

Ebel, B. A. (2012). Wildfire impacts on soil-water retention in the colorado front range, united states. *Water Resources Research*, 48(12).

Eckdahl, J. A., Kristensen, J. A., and Metcalfe, D. B. (2023). Climate and forest properties explain wildfire impact on microbial community and nutrient mobilization in boreal soil. In *Frontiers in Forests and Global Change*.

Edgeley, C. M. and Paveglio, T. B. (2017). Community recovery and assistance following large wildfires: The case of the carlton complex fire. *International journal of disaster risk reduction*, 25:137–146.

Eisenman, D., McCaffrey, S., Donatello, I., and Marshal, G. (2015). An ecosystems and vulnerable populations perspective on solastalgia and psychological distress after a wildfire. *EcoHealth*, 12:602–610.

Elfring, C. (1989). Yellowstone: Fire storm over fire management. *BioScience*, 39(10).

Elliott, K. J., Knoepp, J. D., Vose, J. M., and Jackson, W. A. (2013). Interacting effects of wildfire severity and liming on nutrient cycling in a southern appalachian wilderness area. *Plant and Soil*, 366:165–183.

Emmerton, C. A., Cooke, C. A., Hustins, S., Silins, U., Emelko, M. B., Lewis, T., Kruk, M. K., Taube, N., Zhu, D., Jackson, B., et al. (2020). Severe western canadian wildfire affects water quality even at large basin scales. *Water Research*, 183:116071.

Evangeliou, N. and Eckhardt, S. (2020). Uncovering transport, deposition and impact of radionuclides released after the early spring 2020 wildfires in the chernobyl exclusion zone. *Scientific Reports*, 10(1):10655.

Faccenda, K. and Daehler, C. C. (2022). A screening system to predict wildfire risk of invasive plants. *Biological Invasions*, 24(2):575–589.

Fang, T., Hwang, B. C., Kapur, S., Hopstock, K. S., Wei, J., Nguyen, V., Nizkorodov, S. A., and Shiraiwa, M. (2023). Wildfire particulate matter as a source of environmentally persistent free radicals and reactive oxygen species. *Environmental Science: Atmospheres*, 3(3):581–594.

Fedosov, S., Nikolay, V., Lazarev, A., Malichenko, V., and Toropova, M. (2020). The fire-resistant construction for building safety. In *Proceedings of EECE 2019: Energy, Environmental and Construction Engineering 1*, pages 319–327. Springer.

Fernández, I., Cabaneiro, A., and Carballas, T. (1997). Organic matter changes immediately after a wildfire in an atlantic forest soil and comparison with laboratory soil heating. *Soil biology and Biochemistry*, 29(1):1–11.

Ferreira, L. M., Coimbra, A. P., and de Almeida, A. T. (2020). Autonomous system for wildfire and forest fire early detection and control. *Inventions*, 5(3):41.

Finlay, S. E., Moffat, A., Gazzard, R., Baker, D., and Murray, V. (2012). Health impacts of wildfires. *PLoS currents*, 4.

Finney, M. A. (1998). *FARSITE, Fire Area Simulator–model development and evaluation*. Number 4. US Department of Agriculture, Forest Service, Rocky Mountain Research Station.

Flematti, G. R., Merritt, D. J., Piggott, M. J., Trengove, R. D., Smith, S. M., Dixon, K. W., and Ghisalberti, E. L. (2011). Burning vegetation produces cyanohydrins that liberate cyanide and stimulate seed germination. *Nature Communications*, 2(1):360.

Francos, M. and Úbeda, X. (2021). Prescribed fire management. *Current Opinion in Environmental Science & Health*, 21:100250.

Fraser, A. M., Chester, M. V., and Underwood, B. S. (2022). Wildfire risk, post-fire debris flows, and transportation infrastructure vulnerability. *Sustainable and Resilient Infrastructure*, 7(3):188–200.

Freeman, J. J. and Hayes, E. P. (1987). The metabolism of acetonitrile to cyanide by isolated rat hepatocytes. *Toxicological Sciences*, 8(2):263–271.

Fried, J. S. and Fried, B. D. (1996). Simulating wildfire containment with realistic tactics. *Forest Science*, 42(3):267–281.

Fugate, J. S., Maginnis, M. A., Akafuah, N. K., Saito, K., Finney, M. A., and Forthofer, J. (2014). *A focused analysis on lean fire management systems*, chapter 3. Coimbra University Press.

Gao, Z., Fu, Y., Li, Y., Liu, J., Chen, N., and Zhang, X. (2012). Trends of streamflow, sediment load and their dynamic relation for the catchments in the middle reaches of the yellow river over the past five decades. *Hydrology and Earth System Sciences*, 16(9):3219–3231.

Garver, H. L. and Molander, E. G. (1962). *Fire-resistant Construction on the Farm*. Number 2070. US Department of Agriculture.

Gatebe, C., Ichoku, C., Poudyal, R., Román, M., and Wilcox, E. (2014). Surface albedo darkening from wildfires in northern sub-saharan africa. *Environmental Research Letters*, 9(6):065003.

Gellman, J., Walls, M., and Wibbenmeyer, M. (2022). Wildfire, smoke, and outdoor recreation in the western united states. *Forest Policy and Economics*, 134:102619.

Giannakidou, S., Radoglou-Grammatikis, P., Lagkas, T., Argyriou, V., Goudos, S., Markakis, E. K., and Sarigiannidis, P. (2024). Leveraging the power of internet of things and artificial intelligence in forest fire prevention, detection, and restoration: A comprehensive survey. *Internet of Things*, 26:101171.

Gibbons, P., Van Bommel, L., Gill, A. M., Cary, G. J., Driscoll, D. A., Bradstock, R. A., Knight, E., Moritz, M. A., Stephens, S. L., and Lindenmayer, D. B. (2012). Land management practices associated with house loss in wildfires. *PloS one*, 7(1):e29212.

Ginzburg, O. and Steinberger, Y. (2012). Effects of forest wildfire on soil microbial-community activity and chemical components on a temporal-seasonal scale. *Plant and soil*, 360:243–257.

Glikson, A. (2013). Fire and human evolution: The deep-time blueprints of the anthropocene. *Anthropocene*, 3:89–92.

Gong, P. and Wang, X. (2021). Forest fires enhance the emission and transport of persistent organic pollutants and polycyclic aromatic hydrocarbons from the central himalaya to the tibetan plateau. *Environmental Science & Technology Letters*, 8(7):498–503.

Gonzalez-Mathiesen, C. and March, A. (2018). Establishing design principles for wildfire resilient urban planning. *Planning Practice & Research*, 33(2):97–119.

González-Pérez, J. A., González-Vila, F. J., Almendros, G., and Knicker, H. (2004). The effect of fire on soil organic matter—a review. *Environment international*, 30(6):855–870.

Gorham, E. (1976). Acid precipitation and its influence upon aquatic ecosystems—an overview. *Water, air, and soil pollution*, 6:457–481.

Gowlett, J. A. (2016). The discovery of fire by humans: a long and convoluted process. *Philosophical Transactions of the Royal Society B: Biological Sciences*, 371(1696):20150164.

Graham, A. (2020). Extinguishing wildfires by saturation water bombing: A new concept. *Journal of Emergency Management (Weston, Mass.)*, 18(1):15–25.

Grant, G., Brenton, J., and Drysdale, D. (2000). Fire suppression by water sprays. *Progress in energy and combustion science*, 26(2):79–130.

Grantz, D., Garner, J., and Johnson, D. (2003). Ecological effects of particulate matter. *Environment international*, 29(2-3):213–239.

Green, J. (1995). An overview of the fire retardant chemicals industry, past—present—future. *Fire and materials*, 19(5):197–204.

Guo, D., Saft, M., Hou, X., Webb, J. A., Hairsine, P. B., and Western, A. W. (2023). How does wildfire and climate variability affect streamflow in forested catchments? a regional study in eastern australia. *Journal of Hydrology*, 625:129979.

Haines, A. L. (1974). *Yellowstone National Park: its exploration and establishment*. US National Park Service.

Haines, T. K., Renner, C. R., and Reams, M. A. (2008). A review of state and local regulation for wildfire mitigation. *The Economics of Forest Disturbances: Wildfires, Storms, and Invasive Species*, pages 273–293.

Hakes, R. S., Caton, S. E., Gorham, D. J., and Gollner, M. J. (2017). A review of pathways for building fire spread in the wildland urban interface part ii: response of components and systems and mitigation strategies in the united states. *Fire technology*, 53:475–515.

Hallema, D. W., Robinne, F.-N., and Bladon, K. D. (2018). Reframing the challenge of global wildfire threats to water supplies. *Earth's Future*, 6(6):772–776.

Halofsky, J. E., Peterson, D. L., and Harvey, B. J. (2020). Changing wildfire, changing forests: the effects of climate change on fire regimes and vegetation in the pacific northwest, usa. *Fire Ecology*, 16(1):1–26.

Han, Y.-T., Na, B.-G., Choi, J.-S., and Min, S.-H. (2017). The development of high pressure long distance fire-fighting hose with phosphorescent performance. *Fire Science and Engineering*, 31(5):63–69.

Harrison, R. (1986). Secondary pollutants. In *Handbook of air pollution analysis*, pages 343–386. Springer.

Harun-Ur-Rashid, M., Imran, A. B., and Susan, M. A. B. H. (2023). Fire-resistant polymeric foams and their applications. In *Polymeric Foams: Applications of Polymeric Foams (Volume 2)*, pages 97–121. ACS Publications.

Hesseln, H. (2018). Wildland fire prevention: a review. *Current Forestry Reports*, 4:178–190.

Heyns, A., Du Plessis, W., Kosch, M., and Hough, G. (2019). Optimisation of tower site locations for camera-based wildfire detection systems. *International journal of wildland fire*, 28(9):651–665.

Hitt, N. P. (2003). Immediate effects of wildfire on stream temperature. *Journal of Freshwater Ecology*, 18(1).

Hoang, D. M., Reid, D., and Lentz, C. W. (2013). Statewide ban on recreational fires resulted in a significant decrease in campfire-related summer burn center admissions. *Journal of Burn Care & Research*, 34(1):74–77.

Hoffmann, W. A. and Jackson, R. B. (2000). Vegetation–climate feedbacks in the conversion of tropical savanna to grassland. *Journal of climate*, 13(9):1593–1602.

Hohner, A. K., Rhoades, C. C., Wilkerson, P., and Rosario-Ortiz, F. L. (2019). Wildfires alter forest watersheds and threaten drinking water quality. *Accounts of Chemical Research*, 52(5):1234–1244.

Holden, S. R., Gutierrez, A., and Treseder, K. K. (2013). Changes in soil fungal communities, extracellular enzyme activities, and litter decomposition across a fire chronosequence in alaskan boreal forests. *Ecosystems*, 16:34–46.

Horrocks, A. R. and Price, D. (2008). *Advances in fire retardant materials*. Elsevier.

Hou, X., Orth, R., et al. (2020). Observational evidence of wildfire-promoting soil moisture anomalies. *Scientific reports*, 10(1):1–8.

Hu, X. and Ntaimo, L. (2009). Integrated simulation and optimization for wildfire containment. *ACM Transactions on Modeling and Computer Simulation (TOMACS)*, 19(4):1–29.

Hua, J., Kumar, K., Khoo, B. C., and Xue, H. (2002). A numerical study of the interaction of water spray with a fire plume. *Fire Safety Journal*, 37(7):631–657.

Huang, X., Ding, K., Liu, J., Wang, Z., Tang, R., Xue, L., Wang, H., Zhang, Q., Tan, Z.-M., Fu, C., et al. (2023). Smoke-weather interaction affects extreme wildfires in diverse coastal regions. *Science*, 379(6631):457–461.

Huffman, E. L., MacDonald, L. H., and Stednick, J. D. (2001). Strength and persistence of fire-induced soil hydrophobicity under ponderosa and lodgepole pine, colorado front range. *Hydrological Processes*, 15(15):2877–2892.

Huggett Jr, R. J., Abt, K. L., and Shepperd, W. (2008). Efficacy of mechanical fuel treatments for reducing wildfire hazard. *Forest Policy and Economics*, 10(6):408–414.

Hunter, S. (1993). The oiling of ics. In *International Oil Spill Conference*, volume 1993, pages 25–30. American Petroleum Institute.

Hyde, K., Dickinson, M. B., Bohrer, G., Calkin, D., Evers, L., Gilbertson-Day, J., Nicolet, T., Ryan, K., and Tague, C. (2012). Research and development supporting risk-based wildfire effects prediction for fuels and fire management: status and needs. *International Journal of Wildland Fire*, 22(1):37–50.

Ice, G. G., Neary, D. G., and Adams, P. W. (2004). Effects of wildfire on soils and watershed processes. *Journal of Forestry*, 102(6):16–20.

Igarashi, Y., Onda, Y., Wakiyama, Y., Konoplev, A., Zheleznyak, M., Lisovyi, H., Laptev, G., Damiyanovich, V., Samoilov, D., Nanba, K., et al. (2020). Impact of wildfire on 137cs and 90sr wash-off in heavily contaminated forests in the chernobyl exclusion zone. *Environmental Pollution*, 259:113764.

Intini, P., Ronchi, E., Gwynne, S., and Benichou, N. (2020). Guidance on design and construction of the built environment against wildland urban interface fire hazard: a review. *Fire technology*, 56(5):1853–1883.

Irwin, J. and Williams, M. (1988). Acid rain: chemistry and transport. *Environmental pollution*, 50(1-2):29–59.

Isaak, D. J., Luce, C. H., Rieman, B. E., Nagel, D. E., Peterson, E. E., Horan, D. L., Parkes, S., and Chandler, G. L. (2010). Effects of climate change and wildfire on stream temperatures and salmonid thermal habitat in a mountain river network. *Ecological Applications*, 20(5):1350–1371.

Jabeen, R., Ahmad, A., and Iqbal, M. (2009). Phytoremediation of heavy metals: physiological and molecular mechanisms. *The Botanical Review*, 75:339–364.

Jackson, S. T. and Hobbs, R. J. (2009). Ecological restoration in the light of ecological history. *science*, 325(5940):567–569.

Jackson, W. T. (1942). The creation of yellowstone national park. *The Mississippi Valley Historical Review*, 29(2):187–206.

Jaffe, D. A. and Wigder, N. L. (2012). Ozone production from wildfires: A critical review. *Atmospheric Environment*, 51:1–10.

Jain, P., Coogan, S. C., Subramanian, S. G., Crowley, M., Taylor, S., and Flannigan, M. D. (2020). A review of machine learning applications in wildfire science and management. *Environmental Reviews*, 28(4):478–505.

Jakes, P. J., Nelson, K., Lang, E., Monroe, M., Agrawal, S., Kruger, L., and Sturtevant, V. (2003). A model for improving community preparedness for wildfire. In *Homeowners, Communities, and Wildfire: Science Findings from the National Fire Plan: Proceedings of the Ninth International Symposium on Society and Resource Management,*

Bloomington, Indiana, June 2-5, 2002, volume 231, page 4. US Department of Agriculture, Forest Service, North Central Research Station.

Jazebi, S., De Leon, F., and Nelson, A. (2019). Review of wildfire management techniques—part ii: Urgent call for investment in research and development of preventative solutions. *IEEE Transactions on Power Delivery*, 35(1):440–450.

Jensen, D., Reager, J. T., Zajic, B., Rousseau, N., Rodell, M., and Hinkley, E. (2018). The sensitivity of us wildfire occurrence to pre-season soil moisture conditions across ecosystems. *Environmental Research Letters*, 13(1):014021.

Jensen, J. and Thompson, S. (2016). The incident command system: a literature review. *Disasters*, 40(1):158–182.

Jepson, W. (2005). A disappearing biome? reconsidering land-cover change in the brazilian savanna. *Geographical Journal*, 171(2):99–111.

Jiménez-Morillo, N. T., Almendros, G., José, M., Jordán, A., Zavala, L. M., Granged, A. J., and González-Pérez, J. A. (2020). Effect of a wildfire and of post-fire restoration actions in the organic matter structure in soil fractions. *Science of the Total Environment*, 728:138715.

Jiménez-Morillo, N. T., Spangenberg, J. E., Miller, A. Z., Jordán, A., Zavala, L. M., González-Vila, F. J., and González-Pérez, J. A. (2017). Wildfire effects on lipid composition and hydrophobicity of bulk soil and soil size fractions under quercus suber cover (sw-spain). *Environmental research*, 159:394–405.

Johnson, D. W., Turner, J., and Kelly, J. (1982). The effects of acid rain on forest nutrient status. *Water Resources Research*, 18(3):449–461.

Johnson, L. (2022). *Paradise: One Town's Struggle to Survive an American Wildfire*. Crown.

Johnson, M., Rew, L. J., Maxwell, B. D., and Sutherland, S. (2006). The role of wildfire in the establishment and range expansion of nonnative plant species into natural areas: a review of current literature. *Bozeman, MT: Montana State University Center for Invasive Plant Management. 80 p.*

Joshi, C. C., Payyavula, J. S., Patel, S., and Alginahi, Y. M. (2024). Ml-based wildfire prediction and detection. In *2024 IEEE 3rd International Conference on Computing and Machine Intelligence (ICMI)*, pages 1–5. IEEE.

Kal'avský, P., Petríček, P., Kelemen, M., Rozenberg, R., Jevčák, J., Tomaško, R., and Mikula, B. (2019). The efficiency of aerial firefighting in varying flying conditions. In *2019 International Conference on Military Technologies (ICMT)*, pages 1–5. IEEE.

Kamiya, K., Ozasa, K., Akiba, S., Niwa, O., Kodama, K., Takamura, N., Zaharieva, E. K., Kimura, Y., and Wakeford, R. (2015). Long-term effects of radiation exposure on health. *The lancet*, 386(9992):469–478.

Keefer, W. R. (1971). *The geologic story of Yellowstone National park*, volume 1347. US Government Printing Office.

Keeley, J. E. (2006). Fire management impacts on invasive plants in the western united states. *Conservation Biology*, 20(2):375–384.

Keller, P. A. and Block, L. G. (1997). Vividness effects: A resource-matching perspective. *Journal of Consumer Research*, 24(3):295–304.

Kelly, J., Ibáñez, T. S., Santín, C., Doerr, S. H., Nilsson, M.-C., Holst, T., Lindroth, A., and Kljun, N. (2021). Boreal forest soil carbon fluxes one year after a wildfire: Effects of burn severity and management. *Global Change Biology*, 27(17):4181–4195.

Kennedy, E. B. (2019). Values in science: lessons from wildfires. *Environmental Communication*, 13(2):276–280.

Kent, D. (2019). *Firescaping: protecting your home with a fire-resistant landscape*. Wilderness Press.

Khaledi, J., Lane, P. N., Nitschke, C. R., and Nyman, P. (2022). Wildfire contribution to streamflow variability across australian temperate zone. *Journal of Hydrology*, 609:127728.

Khosravi, R., Pourghasemi, H. R., Adavoudi, R., Julaie, L., and Wan, H. Y. (2022). A spatially explicit analytical framework to assess wildfire risks on brown bear habitat and corridors in conservation areas. *Fire Ecology*, 18(1):1–14.

Khurana, D., Koli, A., Khatter, K., and Singh, S. (2023). Natural language processing: state of the art, current trends and challenges. *Multimedia tools and applications*, 82(3):3713–3744.

Kieta, K. A., Owens, P. N., Petticrew, E., French, T., Koiter, A. J., and Rutherford, P. (2022). Polycyclic aromatic hydrocarbons in terrestrial and aquatic environments following wildfire: a review. *Environmental Reviews*, 31(1):141–167.

Kilinc, F. S. (2013). *Handbook of fire resistant textiles*. Elsevier.

Kim, M.-K. and Jakus, P. M. (2019). Wildfire, national park visitation, and changes in regional economic activity. *Journal of Outdoor Recreation and Tourism*, 26:34–42.

Kinnaird, M. F. and O'Brien, T. G. (1998). Ecological effects of wildfire on lowland rainforest in sumatra. *Conservation Biology*, 12(5):954–956.

Kishnani, N. (2019). Sustainable development of ecotourism in madhya pradesh: Prospects and problems. *Journal of Tourism*, 20(1):55.

Knicker, H. (2007). How does fire affect the nature and stability of soil organic nitrogen and carbon? a review. *Biogeochemistry*, 85(1):91–118.

Kochanski, A. K., Mallia, D. V., Fearon, M. G., Mandel, J., Souri, A. H., and Brown, T. (2019). Modeling wildfire smoke feedback mechanisms using a coupled fire-atmosphere model with a radiatively active aerosol scheme. *Journal of Geophysical Research: Atmospheres*, 124(16):9099–9116.

Koopmans, E., Fyfe, T., Eadie, M., and Pelletier, C. A. (2020). Exploring prevention and mitigation strategies to reduce the health impacts of occupational exposure to wildfires for wildland firefighters and related personnel: protocol of a scoping study. *Systematic reviews*, 9:1–8.

Köster, K., Aaltonen, H., Berninger, F., Heinonsalo, J., Köster, E., Ribeiro-Kumara, C., Sun, H., Tedersoo, L., Zhou, X., and Pumpanen, J. (2021). Impacts of wildfire on soil microbiome in boreal environments. *Current Opinion in Environmental Science & Health*, 22:100258.

Krueger, E. S., Levi, M. R., Achieng, K. O., Bolten, J. D., Carlson, J., Coops, N. C., Holden, Z. A., Magi, B. I., Rigden, A. J., and Ochsner, T. E. (2022). Using soil moisture information to better understand and predict wildfire danger: a review of recent developments and outstanding questions. *International Journal of Wildland Fire*.

Kutchartt, E., González-Olabarria, J. R., Trasobares, A., de Miguel, S., Cardil, A., Botequim, B., Vassilev, V., Palaiologou, P., Rogai, M., and Pirotti, F. (2023). Fire-res geo-catch: a mobile application to support reliable fuel mapping at a pan-european scale. *iForest-Biogeosciences and Forestry*, 16(5):268.

Lam, S. S., Waugh, C., Peng, W., and Sonne, C. (2020). Wildfire puts koalas at risk of extinction. *Science*, 367(6479):750–750.

Lanctôt, C., Grogan, L. F., Tunstill, K., and Melvin, S. D. (2024). Metabolomic response of striped marsh frog (limnodynastes peronii) tadpoles exposed to the fire retardant phos-chek lc95w. *Comparative Biochemistry and Physiology Part C: Toxicology & Pharmacology*, 276:109786.

Lareau, N. P. and Clements, C. B. (2016). Environmental controls on pyrocumulus and pyrocumulonimbus initiation and development. *Atmospheric Chemistry and Physics*, 16(6):4005–4022.

Leavesley, H. B., Li, L., Prabhakaran, K., Borowitz, J. L., and Isom, G. E. (2008). Interaction of cyanide and nitric oxide with cytochrome c oxidase: implications for acute cyanide toxicity. *Toxicological Sciences*, 101(1):101–111.

Leblon, B. (2001). Forest wildfire hazard monitoring using remote sensing: A review. *Remote Sensing Reviews*, 20(1):1–43.

Lei, P., Marfia, G., Pau, G., and Tse, R. (2018). Can we monitor the natural environment analyzing online social network posts? a literature review. *Online Social Networks and Media*, 5:51–60.

Leikauf, G. D. (2002). Hazardous air pollutants and asthma. *Environmental Health Perspectives*, 110(suppl 4):505–526.

Lemons, R. E., Prichard, S. J., and Kerns, B. K. (2023). Evaluating fireline effectiveness across large wildfire events in north-central washington state. *Fire Ecology*, 19(1):8.

Leonard, S. S., Castranova, V., Chen, B. T., Schwegler-Berry, D., Hoover, M., Piacitelli, C., and Gaughan, D. M. (2007). Particle size-dependent radical generation from wildland fire smoke. *Toxicology*, 236(1-2):103–113.

Lewis, J. S., LeSueur, L., Oakleaf, J., and Rubin, E. S. (2022). Mixed-severity wildfire shapes habitat use of large herbivores and carnivores. *Forest Ecology and Management*, 506:119933.

Li, G., Wang, J., Zheng, Y., and Franklin, M. J. (2016). Crowdsourced data management: A survey. *IEEE Transactions on Knowledge and Data Engineering*, 28(9):2296–2319.

Li, G., Zheng, Y., Fan, J., Wang, J., and Cheng, R. (2017). Crowdsourced data management: Overview and challenges. In *Proceedings of the 2017 ACM international conference on Management of Data*, pages 1711–1716.

Li, H., Speer, J. H., and Thapa, I. (2022). Analyzing resilience in the greater yellowstone ecosystem after the 1988 wildfire in the western us using remote sensing and soil database. *Land*, 11(8):1172.

Liang, Y., Sengupta, D., Campmier, M. J., Lunderberg, D. M., Apte, J. S., and Goldstein, A. H. (2021). Wildfire smoke impacts on indoor air quality assessed using crowdsourced data in california. *Proceedings of the National Academy of Sciences*, 118(36):e2106478118.

Liang, Y., Stamatis, C., Fortner, E. C., Wernis, R. A., Van Rooy, P., Majluf, F., Yacovitch, T. I., Daube, C., Herndon, S. C., Kreisberg, N. M., et al. (2022). Emissions of organic compounds from western us wildfires and their near-fire transformations. *Atmospheric Chemistry and Physics*, 22(15):9877–9893.

Likens, G. E., Driscoll, C. T., and Buso, D. C. (1996). Long-term effects of acid rain: response and recovery of a forest ecosystem. *Science*, 272(5259):244–246.

Lindaas, J., Pollack, I. B., Garofalo, L. A., Pothier, M. A., Farmer, D. K., Kreidenweis, S. M., Campos, T. L., Flocke, F., Weinheimer, A. J., Montzka, D. D., et al. (2021). Emissions of reactive nitrogen from western us wildfires during summer 2018. *Journal of Geophysical Research: Atmospheres*, 126(2):e2020JD032657.

Lindell, B. (1987). Radiation and health. *Bulletin of the World Health Organization*, 65(2):139.

Lindenmayer, D. and Possingham, H. P. (1995). Modelling the impacts of wildfire on the viability of metapopulations of the endangered australian species of arboreal marsupial, leadbeater's possum. *Forest Ecology and Management*, 74(1-3):197–222.

Liu, C.-C., Portmann, R. W., Liu, S., Rosenlof, K. H., Peng, Y., and Yu, P. (2022).

Significant effective radiative forcing of stratospheric wildfire smoke. *Geophysical Research Letters*, 49(17):e2022GL100175.

Liu, X., Huey, L. G., Yokelson, R. J., Selimovic, V., Simpson, I. J., Müller, M., Jimenez, J. L., Campuzano-Jost, P., Beyersdorf, A. J., Blake, D. R., et al. (2017). Airborne measurements of western us wildfire emissions: Comparison with prescribed burning and air quality implications. *Journal of Geophysical Research: Atmospheres*, 122(11):6108–6129.

Liu, Y., Goodrick, S., and Heilman, W. (2014). Wildland fire emissions, carbon, and climate: Wildfire–climate interactions. *Forest Ecology and Management*, 317:80–96.

Liu, Y., Stanturf, J., and Goodrick, S. (2010). Trends in global wildfire potential in a changing climate. *Forest ecology and management*, 259(4):685–697.

Loh, M. M., Levy, J. I., Spengler, J. D., Houseman, E. A., and Bennett, D. H. (2007). Ranking cancer risks of organic hazardous air pollutants in the united states. *Environmental health perspectives*, 115(8):1160–1168.

Long, J. W., Quinn-Davidson, L. N., and Skinner, C. N. (2014). Science synthesis to support socioecological resilience in the sierra nevada and southern cascade range.

Lourenço, M., Oliveira, L. B., Oliveira, J. P., Mora, A., Oliveira, H., and Santos, R. (2021). An integrated decision support system for improving wildfire suppression management. *ISPRS International Journal of Geo-Information*, 10(8):497.

Lozier, J. (1976). Volunteer fire departments and community mobilization. *Human Organization*, 35(4):345–354.

Lu, J., Zhou, T., Wu, C., and Ou, Y. (2023). Dropping fire retardants by helicopter and its application to wildfire prevention near electrical transmission lines. *Fire*, 6(5):176.

Lueck, D. (2013). Economics and the organization of wildfire suppression. In *Wildfire Policy*, pages 71–88. Routledge.

Lyon, R. E. et al. (1996). Fire-resistant materials: research overview. Technical report, U.S. Department of Transportation Federal Aviation Administration.

Ma, Z., Li, L., Mao, Y., Wang, Y., Patsy, O. G., Bensi, M. T., Hemphill, L., and Baecher, G. B. (2024). Surveying the use of social media data and natural language processing techniques to investigate natural disasters. *Natural Hazards Review*, 25(4):03124003.

Maia, P., Pausas, J., Arcenegui, V., Guerrero, C., Pérez-Bejarano, A., Mataix-Solera, J., Varela, M., Fernandes, I., Pedrosa, E., and Keizer, J. (2012). Wildfire effects on the soil seed bank of a maritime pine stand—the importance of fire severity. *Geoderma*, 191:80–88.

Marquette, D. (2023). Integrating climate and wildfire education in the classroom: Development and implementation of k-12 place-based wildfire educational modules. Master's thesis, Boise State University.

Martell, D. L. (2015). A review of recent forest and wildland fire management decision support systems research. *Current Forestry Reports*, 1:128–137.

Martinsson, J., Pédehontaa-Hiaa, G., Malmborg, V., Madsen, D., and Rääf, C. (2021). Experimental wildfire induced mobility of radiocesium in a boreal forest environment. *Science of The Total Environment*, 792:148310.

Masson, O., Romanenko, O., Saunier, O., Kirieiev, S., Protsak, V., Laptev, G., Voitsekhovych, O., Durand, V., Coppin, F., Steinhauser, G., et al. (2021). Europe-wide atmospheric radionuclide dispersion by unprecedented wildfires in the chernobyl exclusion zone, april 2020. *Environmental Science & Technology*, 55(20):13834–13848.

Matta, K. (2020). Use of aviation technology in forest fire fighting in slovakia. In *Wood & Fire Safety: Proceedings of the 9th International Conference on Wood & Fire Safety 2020 9*, pages 386–393. Springer.

Mawhinney, J., Dlugogorski, B., and Kim, A. (1994). A closer look at the fire extinguishing properties of water mist. *Fire Safety Science*, 4(1):47–60.

McAllister, S., Zimmerman, T., McHugh, C., and Forthofer, J. (2020). Incident command system. In *Encyclopedia of Wildfires and Wildland-Urban Interface (WUI) Fires*, pages 699–707. Springer.

McConnell, K., Whitaker, S., Fussell, E., DeWaard, J., Price, K., and Curtis, K. (2021). Effects of wildfire destruction on migration, consumer credit, and financial distress.

McCormick, P. S. and Anrens, J. (1998). Visualization of wildfire simulations. *IEEE Computer Graphics and Applications*, 18(2):17–19.

McEwen, B. J. (2018). Strangulation, suffocation, and asphyxia. *Veterinary Forensic Pathology, Volume 1*, pages 129–148.

McGarigal, K., Cushman, S. A., and Regan, C. (2005). *Quantifying terrestrial habitat loss and fragmentation: a protocol*. University of Massachusetts, Department of Natural Resources Conservation.

McKenzie, D., Gedalof, Z., Peterson, D. L., and Mote, P. (2004). Climatic change, wildfire, and conservation. *Conservation biology*, 18(4):890–902.

McKinty, J. (1958). The use of explosives as a practical means of disposing of snags in fire-fighting operations. *Australian Forestry*, 22(1):30–35.

McLennan, J., Holgate, A. M., Omodei, M. M., and Wearing, A. J. (2006). Decision making effectiveness in wildfire incident management teams. *Journal of Contingencies and Crisis Management*, 14(1):27–37.

McWethy, D. B., Schoennagel, T., Higuera, P. E., Krawchuk, M., Harvey, B. J., Metcalf, E. C., Schultz, C., Miller, C., Metcalf, A. L., Buma, B., et al. (2019). Rethinking resilience to wildfire. *Nature Sustainability*, 2(9):797–804.

Meador, M. R. and Carlisle, D. M. (2007). Quantifying tolerance indicator values for common stream fish species of the united states. *Ecological indicators*, 7(2):329–338.

Mehaffey, M., Tankersley Jr, R., Miller, L., and Smith, E. (2009). Evaluating habitat vulnerability to hazardous air pollutants in the southeastern united states. *Integrated Environmental Assessment and Management*, 5(1):150–157.

Merino, A., Fonturbel, M. T., Fernández, C., Chávez-Vergara, B., García-Oliva, F., and Vega, J. A. (2018). Inferring changes in soil organic matter in post-wildfire soil burn severity levels in a temperate climate. *Science of the Total Environment*, 627:622–632.

Mihalus, S., Galway, L. P., Robinson, L. W., Duckert, D., and Parenteau, D. (2023). Wildfire management and evacuation in indigenous communities in canada and the united states: A scoping review. *International Journal of Disaster Risk Reduction*, page 104170.

Miller, J. R. and Hobbs, R. J. (2007). Habitat restoration—do we know what we're doing? *Restoration ecology*, 15(3):382–390.

Miller, R. K., Field, C. B., and Mach, K. J. (2020). Barriers and enablers for prescribed burns for wildfire management in california. *Nature Sustainability*, 3(2):101–109.

Minshall, G. W., Brock, J. T., and Varley, J. D. (1989). Wildfires and yellowstone's stream ecosystems. *BioScience*, 39(10):707–715.

Moayedi, H. and Khasmakhi, M. A. S. A. (2023). Wildfire susceptibility mapping using two empowered machine learning algorithms. *Stochastic Environmental Research and Risk Assessment*, 37(1):49–72.

Mockrin, M. H., Fishler, H. K., and Stewart, S. I. (2020). After the fire: Perceptions of land use planning to reduce wildfire risk in eight communities across the united states. *International Journal of Disaster Risk Reduction*, 45:101444.

Mohapatra, A. and Trinh, T. (2022). Early wildfire detection technologies in practice—a review. *Sustainability*, 14(19):12270.

Molina, D., Martínez, E., García, D., and Grillo, F. (2007). Wildland fire analyst: a needed job position and the required education and training. In *IV International Wildland Fire Conference, Seville, Spain http://www. eufirelab. org/toolbox2/library/upload/2403. pdf*.

Monedero, S., Ramirez, J., and Cardil, A. (2019). Predicting fire spread and behaviour on the fireline. wildfire analyst pocket: A mobile app for wildland fire prediction. *Ecological Modelling*, 392:103–107.

Moody, J. A. and Martin, D. A. (2001). Initial hydrologic and geomorphic response following a wildfire in the colorado front range. *Earth Surface Processes and Landforms: The Journal of the British Geomorphological Research Group*, 26(10):1049–1070.

Moody, J. A. and Martin, D. A. (2004). Wildfire impacts on reservoir sedimentation in the western united states. In *Proceedings of the Ninth International Symposium on River Sedimentation*, pages 1095–1102. Citeseer.

Moody, J. A., Martin, D. A., Haire, S. L., and Kinner, D. A. (2008). Linking runoff response to burn severity after a wildfire. *Hydrological Processes: An International Journal*, 22(13):2063–2074.

Moore-Gough, C., Gough, E., and Major, U. H. (2001). Fire-resistant plants for montana landscapes. *Montana State University extension MT200101AG, Reviewed.*

Moritz, M. A., Morais, M. E., Summerell, L. A., Carlson, J., and Doyle, J. (2005). Wildfires, complexity, and highly optimized tolerance. *Proceedings of the National Academy of Sciences*, 102(50):17912–17917.

Muthusaravanan, S., Sivarajasekar, N., Vivek, J., Paramasivan, T., Naushad, M., Prakashmaran, J., Gayathri, V., and Al-Duaij, O. K. (2018). Phytoremediation of heavy metals: mechanisms, methods and enhancements. *Environmental chemistry letters*, 16:1339–1359.

Naganathan, H., Seshasayee, S. P., Kim, J., Chong, W. K., and Chou, J.-S. (2016). Wildfire predictions: Determining reliable models using fused dataset. *Global Journal of Computer Science and Technology*, 16:35–46.

Narciso, D., Melo, M., Raposo, J. V., Cunha, J., and Bessa, M. (2020). Virtual reality in training: an experimental study with firefighters. *Multimedia tools and applications*, 79:6227–6245.

NASA (2019). Yellowstone fire: Then and now.

Nath, L. (2000). *Conservation and management of the tiger, Panthera tigris tigris, in Bandhavgarh National Park, India*. PhD thesis, University of Oxford.

Navarro, K. M., Butler, C. R., Fent, K., Toennis, C., Sammons, D., Ramirez-Cardenas, A., Clark, K. A., Byrne, D. C., Graydon, P. S., Hale, C. R., et al. (2022). The wildland firefighter exposure and health effect (wffehe) study: Rationale, design, and methods of a repeated-measures study. *Annals of work exposures and health*, 66(6):714–727.

Nayebi, M., Marbouti, M., Quapp, R., Maurer, F., and Ruhe, G. (2017). Crowdsourced exploration of mobile app features: A case study of the fort mcmurray wildfire. In *2017 IEEE/ACM 39th International Conference on Software Engineering: Software Engineering in Society Track (ICSE-SEIS)*, pages 57–66. IEEE.

Newland, M. J., Ren, Y., McGillen, M. R., Michelat, L., Daële, V., and Mellouki, A. (2022). No 3 chemistry of wildfire emissions: A kinetic study of the gas-phase reactions of furans with the no 3 radical. *Atmospheric Chemistry and Physics*, 22(3):1761–1772.

Nkwabi, A., SINCLAIR, A. R. E., Metzger, K., and Mduma, S. (2011). Disturbance, species loss and compensation: Wildfire and grazing effects on the avian community and its food supply in the serengeti ecosystem, tanzania. *Austral Ecology*, 36(4):403–412.

Nyman, P., Sheridan, G. J., Smith, H. G., and Lane, P. N. (2014). Modeling the effects of surface storage, macropore flow and water repellency on infiltration after wildfire. *Journal of Hydrology*, 513:301–313.

Oliveira, S., Rocha, J., and Sá, A. (2021). Wildfire risk modeling. *Current Opinion in Environmental Science & Health*, 23:100274.

Olsen, C. S. (2007). *Citizen-agency interactions in planning and decisionmaking after large fires*. US Department of Agriculture, Forest Service, Pacific Northwest Research Station.

Ondei, S., Price, O. F., and Bowman, D. M. (2024). Garden design can reduce wildfire risk and drive more sustainable co-existence with wildfire. *npj Natural Hazards*, 1(1):18.

Or, D., Furtak-Cole, E., Berli, M., Shillito, R., Ebrahimian, H., Vahdat-Aboueshagh, H., and McKenna, S. A. (2023). Review of wildfire modeling considering effects on land surfaces. *Earth-Science Reviews*, 245:104569.

Organization, W. H. et al. (1993). *Acetonitrile*. World Health Organization.

Ortíz-Rodríguez, A. J., Muñoz-Robles, C., and Borselli, L. (2019). Changes in connectivity and hydrological efficiency following wildland fires in sierra madre oriental, mexico. *Science of the Total Environment*, 655:112–128.

Overton, C. T., Lorenz, A. A., James, E. P., Ahmadov, R., Eadie, J. M., Mcduie, F., Petrie, M. J., Nicolai, C. A., Weaver, M. L., Skalos, D. A., et al. (2022). Megafires and thick smoke portend big problems for migratory birds. *Ecology*, 103(1).

Page, W. G. and Butler, B. W. (2019). Assessing wildland firefighter entrapment. *Fire Management Today. 77 (3): 16-19.*, 77(3):16–19.

Pais, C., Miranda, A., Carrasco, J., and Shen, Z.-J. M. (2021). Deep fire topology: Understanding the role of landscape spatial patterns in wildfire occurrence using artificial intelligence. *Environmental Modelling & Software*, 143:105122.

Palinkas, L. A. and Wong, M. (2020). Global climate change and mental health. *Current opinion in psychology*, 32:12–16.

Park, H., Park, J., Lin, S.-H., and Boorady, L. M. (2014). Assessment of firefighters' needs for personal protective equipment. *Fashion and Textiles*, 1:1–13.

Parker, R., Vitalis, A., Walker, R., Riley, D., and Pearce, H. G. (2017). Measuring wildland fire fighter performance with wearable technology. *Applied ergonomics*, 59:34–44.

Parr, C. L., Te Beest, M., and Stevens, N. (2024). Conflation of reforestation with restoration is widespread. *Science*, 383(6684):698–701.

Pastro, L. A., Dickman, C. R., and Letnic, M. (2011). Burning for biodiversity or burning biodiversity? prescribed burn vs. wildfire impacts on plants, lizards, and mammals. *Ecological Applications*, 21(8):3238–3253.

Pati, P. K., Kaushik, P., Malasiya, D., Ray, T., Khan, M., and Khare, P. (2024). Impacts of forest fire frequency on structure and composition of tropical moist deciduous forest communities of bandhavgarh tiger reserve, central india. *Trees, Forests and People*, 15:100489.

Paveglio, T. B., Brenkert-Smith, H., Hall, T., and Smith, A. M. (2015). Understanding social impact from wildfires: Advancing means for assessment. *International Journal of Wildland Fire*, 24(2):212–224.

Pearson, S. M., Turner, M. G., Wallace, L. L., and Romme, W. H. (1995). Winter habitat use by large ungulates following fire in northern yellowstone national park. *Ecological Applications*, 5(3):744–755.

Pellegrini, A. F., Harden, J., Georgiou, K., Hemes, K. S., Malhotra, A., Nolan, C. J., and Jackson, R. B. (2022). Fire effects on the persistence of soil organic matter and long-term carbon storage. *Nature Geoscience*, 15(1):5–13.

Peng, Q., Palm, B. B., Fredrickson, C. D., Lee, B. H., Hall, S. R., Ullmann, K., Campos, T., Weinheimer, A. J., Apel, E. C., Flocke, F., et al. (2021). Observations and modeling of no x photochemistry and fate in fresh wildfire plumes. *ACS Earth and Space Chemistry*, 5(10):2652–2667.

Pereira, J., Mendes, J., Júnior, J. S., Viegas, C., and Paulo, J. R. (2022). A review of genetic algorithm approaches for wildfire spread prediction calibration. *Mathematics*, 10(3):300.

Perkins, K. B. (1987). Volunteer fire departments: Community integration, autonomy and survival. *Human Organization*, 46(4):342–348.

Perrakis, D. D., Lanoville, R. A., Taylor, S. W., and Hicks, D. (2014). Modeling wildfire spread in mountain pine beetle-affected forest stands, british columbia, canada. *Fire Ecology*, 10:10–35.

Peterson, D. W., Dodson, E. K., and Harrod, R. J. (2023). Snag decomposition following stand-replacing wildfires alters wildlife habitat use and surface woody fuels through time. *Ecosphere*.

Peyton, T. O., Steele, R., Steele, T. O., and Mabey, W. R. (1978). *Carbon disulfide, carbonyl sulfide: Literature review and environmental assessment*. Environmental Protection Agency, Office of Health and Ecological Effects.

Pitts Jr, J. and Finlayson, B. (1975). Mechanisms of photochemical air pollution. *Angewandte Chemie International Edition in English*, 14(1):1–15.

Plucinski, M. P. (2019a). Contain and control: wildfire suppression effectiveness at incidents and across landscapes. *Current Forestry Reports*, 5:20–40.

Plucinski, M. P. (2019b). Fighting flames and forging firelines: Wildfire suppression effectiveness at the fire edge. *Current Forestry Reports*, 5:1–19.

Plucinski, M. P. and Pastor, E. (2013). Criteria and methodology for evaluating aerial wildfire suppression. *International Journal of Wildland Fire*, 22(8):1144–1154.

Potvin, D. A., Parris, K. M., Smith Date, K. L., Keely, C. C., Bray, R. D., Hale, J., Hunjan, S., Austin, J. J., and Melville, J. (2017). Genetic erosion and escalating extinction risk in frogs with increasing wildfire frequency. *Journal of Applied Ecology*, 54(3):945–954.

Prestemon, J. P., Butry, D. T., Abt, K. L., and Sutphen, R. (2010). Net benefits of wildfire prevention education efforts. *Forest Science*, 56(2):181–192.

Prins, E. M., Schmetz, J., Flynn, L. P., Hillger, D. W., and Feltz, J. M. (2001). An overview of diurnal active fire monitoring using a suite of international geostationary satellites. *Global and Regional Vegetation Monitoring from Space: Planning a Coordinated International Effort*, 1:145–170.

Prosser, I. P. and Williams, L. (1998). The effect of wildfire on runoff and erosion in native eucalyptus forest. *Hydrological processes*, 12(2):251–265.

Prudencio, L., Choi, R., Esplin, E., Ge, M., Gillard, N., Haight, J., Belmont, P., and Flint, C. (2018). The impacts of wildfire characteristics and employment on the adaptive management strategies in the intermountain west. *Fire*, 1(3):46.

Qing Li, Q., Loganath, A., Seng Chong, Y., Tan, J., and Philip Obbard, J. (2006). Persistent organic pollutants and adverse health effects in humans. *Journal of Toxicology and Environmental Health, Part A*, 69(21):1987–2005.

Rahman, E. U., Khan, M. A., Algarni, F., Zhang, Y., Irfan Uddin, M., Ullah, I., and Ahmad, H. I. (2021). Computer vision-based wildfire smoke detection using uavs. *Mathematical problems in engineering*, 2021(1):9977939.

Rai, P. K. (2016). Impacts of particulate matter pollution on plants: Implications for environmental biomonitoring. *Ecotoxicology and environmental safety*, 129:120–136.

Rains, M. T. and Hubbard, J. (2002). Protecting communities through the national fire plan. *Fire Management Today*, 62(2):4–12.

Rake, E. L. and Njå, O. (2009). Perceptions and performances of experienced incident commanders. *Journal of Risk Research*, 12(5):665–685.

Reis, V. and Neves, C. (2019). Application of virtual reality simulation in firefighter training for the development of decision-making competences. In *2019 International Symposium on Computers in Education (SIIE)*, pages 1–6. IEEE.

Restaino, C., Eusden, S., and Kay, M. (2024). Taking the next step in wildfire education: integrating multiple knowledge forms into co-produced high school fire science curricula. *Fire Ecology*, 20(1):62.

Rew, L. J. and Johnson, M. P. (2010). Reviewing the role of wildfire on the occurrence and spread of invasive plant species in wildland areas of the intermountain western united states. *Invasive Plant Science and Management*, 3(4):347–364.

Rey, D. M., Briggs, M. A., Walvoord, M. A., and Ebel, B. A. (2023). Wildfire-induced shifts in groundwater discharge to streams identified with paired air and stream water temperature analyses. *Journal of Hydrology*, 619:129272.

Rhoades, C. C., Entwistle, D., and Butler, D. (2011). The influence of wildfire extent and severity on streamwater chemistry, sediment and temperature following the hayman fire, coloradoa. *International Journal of Wildland Fire*, 20(3):430–442.

Rhoades, C. C., Nunes, J. P., Silins, U., and Doerr, S. H. (2019). The influence of wildfire on water quality and watershed processes: New insights and remaining challenges. *International Journal of Wildland Fire*, 28(10):721–725.

Richardson, D., Black, A. S., Irving, D., Matear, R. J., Monselesan, D. P., Risbey, J. S., Squire, D. T., and Tozer, C. R. (2022). Global increase in wildfire potential from compound fire weather and drought. *NPJ climate and atmospheric science*, 5(1):23.

Robichaud, P. R. (2009). Post-fire stabilization and rehabilitation. *Fire effects on soils and restoration strategies*, 5:299–320.

Robinne, F.-N., Bladon, K. D., Miller, C., Parisien, M.-A., Mathieu, J., and Flannigan, M. D. (2018). A spatial evaluation of global wildfire-water risks to human and natural systems. *Science of the Total Environment*, 610:1193–1206.

Robinne, F.-N. and Secretariat, F. (2021). Impacts of disasters on forests, in particular forest fires. *UNFFS Background paper*.

Rodríguez, J., González-Pérez, J. A., Turmero, A., Hernández, M., Ball, A. S., González-Vila, F. J., and Arias, M. E. (2017). Wildfire effects on the microbial activity and diversity in a mediterranean forest soil. *Catena*, 158:82–88.

Romme, W. H., Boyce, M. S., Gresswell, R., Merrill, E. H., Minshall, G. W., Whitlock, C., and Turner, M. G. (2011). Twenty years after the 1988 yellowstone fires: lessons about disturbance and ecosystems. *Ecosystems*, 14:1196–1215.

Romme, W. H. and Despain, D. G. (1989). Historical perspective on the yellowstone fires of 1988. *BioScience*, 39(10):695–699.

Romme, W. H. and Turner, M. G. (2004). Ten years after the 1988 yellowstone fires:

is restoration needed. *After the fires: the ecology of change in Yellowstone National Park*, pages 318–361.

Rosenfeld, D., Fromm, M., Trentmann, J., Luderer, G., Andreae, M., and Servranckx, R. (2007). The chisholm firestorm: observed microstructure, precipitation and lightning activity of a pyro-cumulonimbus. *Atmospheric Chemistry and Physics*, 7(3):645–659.

Rouet-Leduc, J., Pe'er, G., Moreira, F., Bonn, A., Helmer, W., Shahsavan Zadeh, S. A., Zizka, A., and van der Plas, F. (2021). Effects of large herbivores on fire regimes and wildfire mitigation. *Journal of Applied Ecology*, 58(12):2690–2702.

Rouse, J. W., Haas, R. H., Schell, J. A., Deering, D. W., et al. (1974). Monitoring vegetation systems in the great plains with erts. *NASA Spec. Publ*, 351(1):309.

Rubright, S. L. M., Pearce, L. L., and Peterson, J. (2017). Environmental toxicology of hydrogen sulfide. *Nitric Oxide*, 71:1–13.

Sanderfoot, O. V., Bassing, S. B., Brusa, J. L., Emmet, R. L., Gillman, S. J., Swift, K., and Gardner, B. (2021). A review of the effects of wildfire smoke on the health and behavior of wildlife. *Environmental Research Letters*, 16.

Santelices, R., Ariza, A. C., Flores, P. S., and Cerrillo, R. M. N. (2022). Assessment of a wildfire in the remaining nothofagus alessandrii forests, an endangered species of chile, based on satellite sentinel-2 images. *Ciencia e investigación agraria: revista latinoamericana de ciencias de la agricultura*, 49(2):85–96.

Santos, G., Marques, R., Ribeiro, J., Moreira, A., Fernandes, P., Silva, M., Fonseca, A., Miranda, J. M., Campos, J. B., and Neves, S. F. (2022). Firefighting: challenges of smart ppe. *Forests*, 13(8):1319.

Sauvageau, A. and Boghossian, E. (2010). Classification of asphyxia: the need for standardization. *Journal of forensic sciences*, 55(5):1259–1267.

Schindler, D. E. (2017). Warmer climate squeezes aquatic predators out of their preferred habitat. *Proceedings of the National Academy of Sciences*, 114:9764 – 9765.

Schlesinger, R. B. (2007). The health impact of common inorganic components of fine particulate matter (pm2. 5) in ambient air: a critical review. *Inhalation toxicology*, 19(10):811–832.

Scordo, F., Chandra, S., Suenaga, E., Kelson, S. J., Culpepper, J., Scaff, L., Tromboni, F., Caldwell, T. J., Seitz, C., Fiorenza, J. E., Williamson, C. E., Sadro, S., Rose, K. C., and Poulson, S. R. (2021). Smoke from regional wildfires alters lake ecology. *Scientific Reports*, 11.

Sellers, R. E. and Despain, D. G. (1976). Fire management in yellowstone national park. In *Proceedings, Tall Timbers Fire Ecology Conference*, volume 14, pages 99–113.

Shakeriaski, F. and Ghodrat, M. (2022). Challenges and limitation of wearable sensors used in firefighters' protective clothing. *Journal of fire sciences*, 40(3):214–245.

Shakesby, R. A. (2011). Post-wildfire soil erosion in the mediterranean: Review and future research directions. *Earth-Science Reviews*, 105(3-4):71–100.

Shakesby, R. A. and Doerr, S. H. (2006). Wildfire as a hydrological and geomorphological agent. *Earth-Science Reviews*, 74(3-4):269–307.

Shaw, R. (2014). Post disaster recovery: Issues and challenges. *Disaster recovery: Used or misused development opportunity*, pages 1–13.

Shiralipour, H. J., Monroe, M. C., Nelson, K. C., and Payton, M. (2006). Working with neighborhood organizations to promote wildfire preparedness. *The public and wildland fire management: social science findings for managers'.(Ed. SM McCaffrey) USDA Forest Service, Northern Research Station General Technical Report NRS-1*, pages 151–162.

Sigmund, G., Santín, C., Pignitter, M., Tepe, N., Doerr, S. H., and Hofmann, T. (2021). Environmentally persistent free radicals are ubiquitous in wildfire charcoals and remain stable for years. *Communications Earth & Environment*, 2(1):68.

Sillman, S. (2003). Tropospheric ozone and photochemical smog. *Treatise on geochemistry*, 9:612.

Silva, J., Marques, J., Gonçalves, I., Brito, R., Teixeira, S., Teixeira, J., and Alvelos, F. (2022). A systematic review and bibliometric analysis of wildland fire behavior modeling. *Fluids*, 7(12):374.

Šimanský, V. (2015). Changes in soil structure and soil organic matter due to different severities of fire. *Ekológia (Bratislava)*, 34(3):226–234.

Singh, S. S. and Jeganathan, C. (2024). Quantifying forest resilience post forest fire disturbances using time-series satellite data. *Environmental Monitoring and Assessment*, 196(1):26.

Skytt, T., Nielsen, S. N., and Jonsson, B.-G. (2020). Global warming potential and absolute global temperature change potential from carbon dioxide and methane fluxes as indicators of regional sustainability–a case study of jämtland, sweden. *Ecological Indicators*, 110:105831.

Slavkovikj, V., Verstockt, S., Van Hoecke, S., and Van de Walle, R. (2014). Review of wildfire detection using social media. *Fire safety journal*, 68:109–118.

Smith, H. G., Sheridan, G. J., Lane, P. N., Nyman, P., and Haydon, S. (2011). Wildfire effects on water quality in forest catchments: A review with implications for water supply. *Journal of Hydrology*, 396(1-2):170–192.

Steelman, T. A. and Burke, C. A. (2007). Is wildfire policy in the united states sustainable? *Journal of forestry*, 105(2):67–72.

Stephens, K. K., Powers, C. J., Robertson, B. W., Spearing, L. A., Collier, J. C., Tich, K. P., and Smith, W. R. (2023). Building more resilient communities with a wildfire preparedness drill in the us: Individual and community influences and communication practices. *Journal of Contingencies and Crisis Management*, 31(1):26–38.

Stillman, A. N., Siegel, R. B., Wilkerson, R. L., Johnson, M., Howell, C. A., and Tingley, M. W. (2019). Nest site selection and nest survival of black-backed woodpeckers after wildfire. *The Condor*, 121:1 – 13.

Stocks, B. J., Lynham, T., Lawson, B., Alexander, M., Wagner, C. V., McAlpine, R., and Dube, D. (1989). Canadian forest fire danger rating system: an overview. *The Forestry Chronicle*, 65(4):258–265.

Stone, R., Augustine, J., Dutton, E., O'Neill, N., and Saha, A. (2011). Empirical determinations of the longwave and shortwave radiative forcing efficiencies of wildfire smoke. *Journal of Geophysical Research: Atmospheres*, 116(D12).

Stuart-Smith, K. A., Adams, I. T., and Larsen, K. W. (2002). Songbird communities in a pyrogenic habitat mosaic. *International Journal of Wildland Fire*, 11:75–84.

Sturtevant, V. and McCaffrey, S. (2006). Encouraging wildland fire preparedness: Lessons learned from three wildfire education programs. *The public and wildland fire management: Social science findings for managers*, pages 125–136.

Sweeney, K., Dittrich, R., Moffat, S., Power, C., and Kline, J. D. (2023). Estimating the economic value of carbon losses from wildfires using publicly available data sources: Eagle creek fire, oregon 2017. *Fire Ecology*, 19(1):55.

Syphard, A. D., Brennan, T. J., and Keeley, J. E. (2014). The role of defensible space for residential structure protection during wildfires. *International Journal of Wildland Fire*, 23(8):1165–1175.

Taylor, D. L. (1973). Some ecological implications of forest fire control in yellowstone national park, wyoming. *Ecology*, 54(6):1394–1396.

Taylor, M. A. (2014). *Jumping fire: A smokejumper's memoir of fighting wildfire*. HMH.

Taylor Jr, C. A. (2006). Targeted grazing to manage fire risk. *Targeted grazing: A natural approach to vegetation management and landscape enhancement*, pages 107–112.

Tedim, F., Leone, V., Amraoui, M., Bouillon, C., Coughlan, M. R., Delogu, G. M., Fernandes, P. M., Ferreira, C., McCaffrey, S., McGee, T. K., et al. (2018). Defining extreme wildfire events: Difficulties, challenges, and impacts. *Fire*, 1(1):9.

Tedim, F., Leone, V., McCaffrey, S., McGee, T. K., Coughlan, M., Correia, F. J., and Magalhães, C. G. (2020a). Safety enhancement in extreme wildfire events. In *Extreme wildfire events and disasters*, pages 91–115. Elsevier.

Tedim, F., McCaffrey, S., Leone, V., Delogu, G. M., Castelnou, M., McGee, T. K., and Aranha, J. (2020b). What can we do differently about the extreme wildfire problem: An overview. *Extreme Wildfire Events and Disasters*, pages 233–263.

Thomas, D., Butry, D., Gilbert, S., Webb, D., Fung, J., et al. (2017). The costs and losses of wildfires. *NIST Special Publication*, 1215(11).

Toledo, T., Marom, I., Grimberg, E., and Bekhor, S. (2018). Analysis of evacuation behavior in a wildfire event. *International journal of disaster risk reduction*, 31:1366–1373.

Tomaz, S., Cui, T., Chen, Y., Sexton, K. G., Roberts, J. M., Warneke, C., Yokelson, R. J., Surratt, J. D., and Turpin, B. J. (2018). Photochemical cloud processing of primary wildfire emissions as a potential source of secondary organic aerosol. *Environmental science & technology*, 52(19):11027–11037.

Tracey, J. A., Rochester, C. J., Hathaway, S. A., Preston, K. L., Syphard, A. D., Vandergast, A. G., Diffendorfer, J. E., Franklin, J., MacKenzie, J. B., Oberbauer, T. A., et al. (2018). Prioritizing conserved areas threatened by wildfire and fragmentation for monitoring and management. *Plos one*, 13(9):e0200203.

Tracy, R. and Ruby, E. (1994). Burned area emergency rehabilitation; structural treatment, site selection criteria, and project requirements. *USDA Forest Service, Burned Area Emergency Response training course (1998), Region*, 5.

Tunstill, K., Grogan, L. F., Morrison, C., McCallum, H., and Lanctôt, C. (2022). Effects of two firefighting chemical formulations, phos-chek lc95w and blaze-tamer380, on striped marsh frog (limodynastes peronii) tadpole survival, growth, development and behaviour. *Aquatic Toxicology*, 252:106326.

Turner, M. G., Hargrove, W. W., Gardner, R. H., and Romme, W. H. (1994). Effects of fire on landscape heterogeneity in yellowstone national park, wyoming. *Journal of Vegetation Science*, 5(5):731–742.

Turner, M. G., Romme, W. H., and Tinker, D. B. (2003). Surprises and lessons from the 1988 yellowstone fires. *Frontiers in Ecology and the Environment*, 1(7):351–358.

Underwood, R. and Held, A. (2012). Wildfire. In *Handbook of Hazards and Disaster Risk Reduction*, pages 281–293. Routledge.

Urbanski, S. (2014). Wildland fire emissions, carbon, and climate: Emission factors. *Forest Ecology and Management*, 317:51–60.

Urbanski, S. P., Hao, W. M., and Baker, S. (2008). Chemical composition of wildland fire emissions. *Developments in environmental science*, 8:79–107.

Vachula, R. S., Nelson, J. R., and Hall, A. G. (2023). The timing of fireworks-caused wildfire ignitions during the 4th of july holiday season. *PLoS one*, 18(9):e0291026.

Van Der Aalst, W. (2016). *Data science in action*. Springer.

Van Wagner, C. E. et al. (1974). *Structure of the Canadian forest fire weather index*, volume 1333. Environment Canada, Forestry Service Ottawa, ON, Canada.

van Wagtendonk, J. W. (1991). The evolution of national park service fire policy. In *Fire and the Environment: Ecological and Cultural Perspectives: Proceedings of an International Symposium, Knoxville, Tennessee, March 20-24, 1990*, volume 69, page 328. Southeastern Forest Experiment Station.

Varela, M., Benito, E., and De Blas, E. (2005). Impact of wildfires on surface water repellency in soils of northwest spain. *Hydrological Processes: An International Journal*, 19(18):3649–3657.

Vashukevich, N. V., Timoshkin, O. A., Samsónov, D. P., Kulikova, N. N., Levasheva, M. V., and Lukhnev, A. (2023). Taiga fire on bolshoy ushkany island as a model case of forest soil transformation and potential source of eutrophication in lake baikal coastal zone. *Limnology and Freshwater Biology*.

Vejmelka, M., Kochanski, A. K., and Mandel, J. (2016). Data assimilation of dead fuel moisture observations from remote automated weather stations. *International Journal of Wildland Fire*, 25(5):558–568.

Visser, S. and Parkinson, D. (1999). Wildfire vs. clearcutting: impacts on ectomycorrhizal and decomposer fungi. *United States Department of Agriculture Forest Service General Technical Report*, pages 114–123.

Wagle, R. and Eakle, T. W. (1979). A controlled burn reduces the impact of a subsequent wildfire in a ponderosa pine vegetation type. *Forest Science*, 25(1):123–129.

Wang, H.-H., Finney, M. A., Song, Z.-L., Wang, Z.-S., and Li, X.-C. (2021). Ecological techniques for wildfire mitigation: Two distinct fuelbreak approaches and their fusion. *Forest Ecology and Management*, 495:119376.

Wang, Q., Ma, T., Hanson, J., and Larranaga, M. (2012). Application of incident command system in emergency response. *Process Safety Progress*, 31(4):402–406.

Ward, T. J. and Lincoln, E. (2006). Concentrations of pm 2.5-associated oc, ec, and pcdd/fs measured during the 2003 wildfire season in missoula, montana. *Environmental monitoring and assessment*, 115:39–50.

Warrick, J. A. and Rubin, D. M. (2007). Suspended-sediment rating curve response to urbanization and wildfire, santa ana river, california. *Journal of Geophysical Research: Earth Surface*, 112(F2).

Westerling, A. L. and Bryant, B. (2008). Climate change and wildfire in california. *Climatic Change*, 87:231–249.

White, S. C. and Hostler, D. (2017). The effect of firefighter protective garments, self-contained breathing apparatus and exertion in the heat on postural sway. *Ergonomics*, 60(8):1137–1145.

Williams, A. P., Abatzoglou, J. T., Gershunov, A., Guzman-Morales, J., Bishop, D. A., Balch, J. K., and Lettenmaier, D. P. (2019). Observed impacts of anthropogenic climate change on wildfire in california. *Earth's Future*, 7(8):892–910.

Witt, S. (1999). Burned area emergency response, baer. *USDA Forest Service. Watershed, Fish*.

Wolf, K. D., Higuera, P. E., Davis, K. T., and Dobrowski, S. Z. (2021). Wildfire impacts on forest microclimate vary with biophysical context. *Ecosphere*, 12.

Woods, S. W., Birkas, A., and Ahl, R. (2007). Spatial variability of soil hydrophobicity after wildfires in montana and colorado. *Geomorphology*, 86(3-4):465–479.

Wooster, M. J., Roberts, G. J., Giglio, L., Roy, D. P., Freeborn, P. H., Boschetti, L., Justice, C., Ichoku, C., Schroeder, W., Davies, D., et al. (2021). Satellite remote sensing of active fires: History and current status, applications and future requirements. *Remote Sensing of Environment*, 267:112694.

Wu, J., Baartman, J. E., and Nunes, J. P. (2021). Comparing the impacts of wildfire and meteorological variability on hydrological and erosion responses in a mediterranean catchment. *Land Degradation & Development*, 32(2):640–653.

Wu, R., Chen, C., Ahmad, S., Volk, J. M., Luca, C., Harris, F. C., and Dascalu, S. M. (2016). A real-time web-based wildfire simulation system. In *IECON 2016-42nd Annual Conference of the IEEE Industrial Electronics Society*, pages 4964–4969. IEEE.

Xanthopoulos, G., Delogu, G. M., Leone, V., Correia, F. J., and Magalhães, C. G. (2020). Firefighting approaches and extreme wildfires. In *Extreme Wildfire Events and Disasters*, pages 117–132. Elsevier.

Xu, Z., Li, J., and Xu, L. (2024). Wildfire risk prediction: A review. *arXiv preprint arXiv:2405.01607*.

Yoder, J., Engle, D., and Fuhlendorf, S. (2004). Liability, incentives, and prescribed fire for ecosystem management. *Frontiers in Ecology and the Environment*, 2(7):361–366.

Young, S., Balluz, L., and Malilay, J. (2004). Natural and technologic hazardous material releases during and after natural disasters: a review. *Science of the total environment*, 322(1-3):3–20.

Yu, A. C., Lopez Hernandez, H., Kim, A. H., Stapleton, L. M., Brand, R. J., Mellor, E. T., Bauer, C. P., McCurdy, G. D., Wolff III, A. J., Chan, D., et al. (2019). Wildfire prevention through prophylactic treatment of high-risk landscapes using viscoelastic retardant fluids. *Proceedings of the National Academy of Sciences*, 116(42):20820–20827.

Zehra, S. N. and Wong, S. D. (2024). Systematic review and research gaps on wildfire evacuations: infrastructure, transportation modes, networks, and planning. *Transportation Planning and Technology*, pages 1–35.

Zhang, F., Zhao, P., Xu, S., Wu, Y., Yang, X., and Zhang, Y. (2020). Integrating multiple factors to optimize watchtower deployment for wildfire detection. *Science of the total environment*, 737:139561.

Zhang, L., Ma, B., Tang, C., Yu, H., Lv, X., Mazza Rodrigues, J. L., Dahlgren, R. A., and Xu, J. (2021). Habitat heterogeneity induced by pyrogenic organic matter in wildfire-perturbed soils mediates bacterial community assembly processes. *The ISME journal*, 15(7):1943–1955.

Zhang, M., Buekens, A., and Li, X. (2017). Dioxins from biomass combustion: an overview. *Waste and biomass valorization*, 8:1–20.

Zhang, Y., Fan, J., Logan, T., Li, Z., and Homeyer, C. R. (2019). Wildfire impact on environmental thermodynamics and severe convective storms. *Geophysical Research Letters*, 46(16):10082–10093.

Zhong, X., Duckham, M., Chong, D., and Tolhurst, K. (2016). Real-time estimation of wildfire perimeters from curated crowdsourcing. *Scientific reports*, 6(1):24206.

Zong, X., Tian, X., and Wang, X. (2021). An optimal firebreak design for the boreal forest of china. *Science of The Total Environment*, 781:146822.

Zope, V., Dadlani, T., Matai, A., Tembhurnikar, P., and Kalani, R. (2020). Iot sensor and deep neural network based wildfire prediction system. In *2020 4th International Conference on Intelligent Computing and Control Systems (ICICCS)*, pages 205–208. IEEE.

Zouhar, K. (2008). *Wildland fire in ecosystems: fire and nonnative invasive plants*. Forest Service.

Zvereva, E. L. and Kozlov, M. V. (2010). Responses of terrestrial arthropods to air pollution: a meta-analysis. *Environmental Science and Pollution Research*, 17:297–311.

Index